Field Manual for Water Quality Monitoring

An Environmental Education Program for Schools

Twelfth Edition

Mark K. Mitchell, M.S.
William B. Stapp, Ph.D.

Contributors:

**Anne Beebe, Lisa Bryce-Lewis, Jim Bull,
Sally Cole-Misch, Mare Cromwell, Shelley
Downes, Thomas Ellis, Joanne Goodwin,
John Henry, Cheryl Koenig, Lisa LaRocque,
Gail Luera, Tom Martin, Susan Maruca, Vincent
Meldrum, Margaret Pennock, Jennifer Punteney,
Joe Rathbun, Kristi Rosema, David Schmidt,
Julie Schultz, Diane Silver, Arjen Wals, Keith
Wheeler, Clancy Wolf, Mark Zankel**

Additional Editing by:

Kevin Bixby

KENDALL/HUNT PUBLISHING COMPANY
4050 Westmark Drive Dubuque, Iowa 52002

All photos by William B. Stapp unless otherwise credited.
Cover photos by William B. Stapp.

This manual is used in many national and international watershed pro-grams linked through the Global Rivers Environmental Education Network (GREEN). For further information regarding this international network contact:

Earth Force
1908 Mt. Vernon Ave. 2nd Floor
Alexandria, VA 22301
Telephone: 703-299-9400
FAX: 703-299-9485
World Wide Web: www.earthforce.org
Email: green@earthforce.org

Printed in the United States of America
10   9   8   7   6   5   4   3   2

Dedication

To the young people changing their communities and caring for our environment now, while developing life long habits of active citizenship and environmental stewardship.

Contents

Preface

This manual has been prepared to assist citizens in the development of attitudes, knowledge, and skills essential in helping to maintain and improve the water quality of our rivers throughout the world.

We depend on our rivers to provide clean and healthy water for recreational usage, municipal water supplies, fish and wildlife habitats, irrigating croplands, scenic vistas, and meeting daily industrial demands. It is for these reasons, and many more, that all citizens need to understand clearly what rivers mean to the quality of our lives.

Improved water quality in our rivers depends on effecting change not only directly, but also indirectly, through state and national standards for safer water usage. It is important that we know which organizations and agencies are concerned with preventing degradation of our waterways and protecting existing benefits.

The instructional model incorporated in this manual encourages the integration of ecological, economic, political, and social disciplines essential to the resolution of critical water quality issues facing our waterways. If we do not assist our citizenry in becoming alert to the subtle but often drastic changes in water quality in our river systems, we will soon lose many of the amenities that our rivers provide.

The authors would like to acknowledge the creative work of the National Sanitation Foundation in devising the Water Quality Index utilized in this manual. We also thank the many aquatic biologists and managers who have contributed background information. Sally Cole-Misch and Jennifer Punteney were particularly helpful during the early stages of the development of this manual. Dr. John Gannon and Dr. Peter Meier, from the school of Public Health, University of Michigan, offered valuable suggestions for improving the manual, as did Orin Gelderloos and Herm Boatin of the University of Michigan-Dearborn Campus.

We would also like to recognize the School of Natural Resources, University of Michigan, for the support provided throughout this project, and the professional assistance of David Allan, Dr. Jonathan Bulkley, Dr. James Crowfoot, Dr. Raymond DeYoung, Dr. Harrison Morton, Dr. Paul Nowak, Dr. David White, Dr. Michael Wiley, and Dr. Jim Wojcik.

We would also like to thank the many undergraduate and graduate students who offered valuable assistance in the laboratory and many full days of work in various high schools, and whose field work and discussion made an invaluable contribution to the development and preparation of this manual.

We would also like to give full credit to the teachers, administrators, and students who participated in the initial stages of the interactive water quality programs at Dexter High School (science teachers Sam Skidmore and Jean Dalton), Huron High School (biology teachers Dale Greiner, Robert Hubbard, and Terry Pokela), Bellevue High School (biology and ecology teacher Marshal Wied), and Lake Michigan Catholic High School (biology teacher Lynda Smith).

We also owe a great deal of gratitude to the teachers and students of the 100 high schools that were instrumental in the development of the Rouge River program. It is also important to recognize those high schools along the Clinton River, Cuyahoga River, Grand River (Canada), Miami River, Saginaw River, Black River, Rio Grande, and St. Clair River, that participated in, and made valuable contributions to, the later stages of the water monitoring program.

Credit should also be given to Joseph Chadborne and Thomas Offutt of the Institute of Environmental Education, Cleveland, Ohio; Jack Byrne of River Watch; Robert Williams of the Southern Illinois Rivers Project, and Tom Murdoch of Adopt-A-Stream for sharing their enthusiasm and experiences working with students in stream monitoring.

We would like to thank Adopt-A-Stream Foundation for giving us permission to use the storm drain stencil activity in Chapter 7. Likewise, we would like to thank the Industrial States Policy Center and the School of Natural Resources (University of Michigan) for granting us permission to use the Home Hazardous Waste Product Survey in Chapter 7.

We would like to acknowledge the work of the toxic assessment team here at the School of Natural Resources and Environment, from whose good work we have borrowed: Patrick Christie, Kee Condict, Andrew Lewis, Emily Tibbott, and Patty West.

It is indeed a pleasure to recognize the extraordinary talent of Shelley Downes, a 14-year-old graphic artist who prepared the aquatic drawings in Chapter 6. Shelley was a student who participated in the Rouge River Interactive Water Quality Program and whose artistic ability came to our attention during the development of the program. Also, Helen Bunch and John Henry prepared the graphics used in Chapter 3 to help clarify the procedural steps for accurately collecting water quality samples.

It is important to recognize the many private and public bodies that have provided innovative ideas and program funds, such as General Motors Corporation, Ford Motor Co., United States Department of Energy, United States Department of Education, National Consortium of Environmental Education and Training, Public Interest Research Group in Michigan, United Nations Environment Programme, Friends of the Rouge, Walpole Island Heritage Center, Michigan Department of Natural Resources, Kellogg Foundation, Dow Chemical of Canada, Michigan Department of Education, the National Science Foundation, the United States Environmental Protection Agency, Trout Unlimited, George Gund Foundation, Huron River Watershed Council, Frank Butt Memorial Foundation, Polysar Rubber Corporation, University of Michigan, the Bullitt Foundation, World Resources Institute, Institute for Global Communications, National Fish and Wildlife Foundation, Northwest Indian Fisheries Commission, President's Council for Sustainable Development, Tides Foundation, Western Washington University, Abernathy, MacGregor & Scanlon, Key Bank, Owens Corning Corporation, Life-Link Foundation-Sweden, Waldumar Nature Center, Bonneville Power Administration, National Science Foundation, U.S. Peace Corps, National Science Teachers Association, Kiwanis International, TERC, Baltic University Programme, and Stream Watch-Australia.

The most recent contributors to the program have been the students, teachers, administrators and government officials associated with the Global Rivers Environmental Education Network (GREEN) in Latin America, Africa, Middle East, Asia, Australia, Europe, New Zealand, Oceania, and North America.

Finally, credit should be given to Gloria and Tara and other members of our families for providing the support that enabled us to prepare this manual.

To the students, teachers and community citizens that use this manual, we would be pleased if you would share with us your experiences and suggestions so that we might incorporate these valuable comments in the next edition.

CHAPTER

Overview

As a guide to measuring water quality, this manual is a step in the long-term process of restoring the health of the world's rivers. But it is also part of an even larger task; namely, improving our educational systems, and creating new ways of thinking about ourselves and how we relate to other life forms on the planet and to nature as a whole.

More and more people are realizing that we need to find better ways to share our world's natural resources with each other and with other species to ensure that the planet can support life in the future. By allowing poison in our rivers, we are slowly drinking it ourselves. Everything we do affects our water. We are beginning to see how important each of us is to making the world a better place to live, no matter what kind of work we do or how intelligent society says we are.

This manual covers two main components of a model of education that we have found to be a useful approach to learning about river systems and many other topics. The first piece of this model, which is called action research, is the data collection component. The objective of this component is to identify community issues to study. In some cases, students will select these issues, and in others, teachers will already have identified some issues that they think students might want to examine.

In river studies, data collection is done in the form of water quality testing, researching land use, and historical analysis. In addition, students make note of the thoughts, feelings, images, sounds, and smells that strike them in their experiences with the river and its surroundings. These informal observations, or sensory impressions, are not only important, but they often motivate us to make the transition to the second component of the action research model: the problem-solving/action-taking component.

For the problem-solving component, students, teachers, and citizens team up to devise actions to raise the quality of our environment. This component links what we study in school to how we live.

As action researchers we are both learners and doers. We not only test rivers to determine the water's health, but we also prescribe "treatment" or courses of action. In addition to the water quality tests, this manual includes several examples of actions that students and other citizens have taken on behalf of their rivers.

The specific skills that this manual is designed to teach you are:

1. to understand the meaning of nine important tests for measuring water quality;
2. to become familiar with important sources of water pollution and ways to help solve those problems;
3. to learn how to run accurate water quality tests; and determine how the tests relate to each other; and
4. to run the tests on the river safely; and to understand what the test results mean in terms of human uses of the river.

There are many ways to look at a river besides simply testing water quality. When we are trying to change something about a river system, all of the things that affect it must be considered. For example, we need to know what flows into the river, which requires looking at how we use the land around it.

We must expand our vision beyond that of a scientist. We need to perceive rivers in the same way as sociologists, economists, artists and politicians might. Whether we recognize it or not, each of us, regardless of occupation or perception, has a role in the decision-making process about how the river is treated. After all, it is our daily activities and land uses that determine what goes into the water.

In short, becoming a student of the world's rivers requires going beyond any one way of thinking. It calls for skills used in language arts, civics, social studies, and other classes, as well as those employed in science, ecology or biology classes.

We hope that the river helps you to find a purpose for your studies that runs through your daily life, much like the river itself flows through and connects your community.

What Does Monitoring the River Accomplish?

It is important to measure the quality of a river over long periods of time to detect changes in a river ecosystem, including the land around the river as well as the water in it. The data you collect about the health of the water can be used to analyze changes over time.

It is also important to collect data at different points so that water quality along the length of the river can be compared. For example, the river may be less healthy where it passes through farmlands than where it passes through upstream forested areas due to the runoff of fertilizer in rainwater.

As in a human body, not all parts of a river are equally healthy all of the time. And just as getting a physical check-up on a regular basis is important to us, checking the river at regular intervals can provide us with the information we need to maintain or improve water quality.

It is important to not only do the tests, but to let other people know about your results. Notifying community residents and officials about research results can lead to increased public awareness and action directed toward improved water management. Personal lifestyle, industry, and government policy all have an impact on water quality.

How Can I Best Use This Manual?

This manual is divided into clearly labeled chapters and sections, listed in the table of contents, so that you can pick and choose among them as needed or as time allows. However, since all of the sections of the manual are interrelated, we recommend reading the entire book if possible.

We have tried to emphasize the ways in which the different water quality tests relate to each other. Throughout the manual you will notice boxed text sections. These are specifically targeted at instructors (teachers or leaders).

Other sections of the manual have been developed based on feedback from river projects around the globe:

➤ The concept of a watershed is discussed in Chapter 2, as well as techniques for measuring water quality, sampling procedures, and precautions necessary for monitoring water quality.

➤ Chapter 3 contains precise test procedures, with the idea that substitute equipment and techniques may be required depending on such factors as your budget, numbers and experience of participants, and the unique demands of your river ecosystem.

➤ Methods for calculating the overall water quality of a section of a river are given in Chapter 4.

➤ Assessment of Toxics using a bioassay in Chapter 5.

➤ River ecology and the many kinds of aquatic organisms (macroinvertebrates) found in rivers are discussed in Chapter 6, as well as models for sampling and monitoring water quality.

➤ Chapter 7 discusses land uses and their relation to water quality. This chapter also contains information on aerial photographs and satellite images, and activities useful for students in learning more about their local rivers and how they can personally influence water quality.

➤ Chapter 8 contains a case study describing how the Rouge River Education Program is handled in an interdisciplinary manner in one watershed.

➤ Chapter 9 discusses the international dimension of water quality monitoring. River studies around the world are accelerating. The Global Rivers Environmental Education Network (GREEN), founded in 1989 at the University of Michigan School of Natural Resources and Environment, has incorporated the principles of existing river projects. In addition, there is a movement underway to spread the worldwide effort of river monitoring to waterways other than rivers, such as lakes, groundwater, and estuaries.

➤ Chapter 10 discusses the opportunity to share information between schools within a watershed, or between watersheds nationally and internationally by way of computer conferencing.

➤ The last section contains a list of literature to supplement the information presented in this manual.

In conclusion, we cannot overemphasize the significance of creativity in planning your river study. All of the river projects that we have had the pleasure of observing or being involved in have been imaginative in their approach. They go beyond books, just as explorations of unknown territories go beyond maps. As students of rivers, we are explorers, too. We invite you to explore your local waterways and seek the most fun, creative, and educational experience possible. We are also responsible for the health of the river. Use what you learn to take an active role in the protection of your watershed.

Introduction

The River and Its Watershed

Have you ever looked into the flowing waters of a stream or river and wondered where it began, or to where it flowed? Rivers have a mysterious, ancient pull on us to try and unlock some of their secrets. Let's begin to describe and study some of the features unique to rivers.

All rivers share flowing water as their commonality.

Rivers may be large or small, but they all follow certain rules of drainage. Streams may begin from underground water coming to the surface (springs), from rainfall or snowmelt, drainage from a wetland, as meltwater from a glacier, or as an outlet of a lake or pond. Depending upon the physical nature of the land, a stream may be crowded with vegetation and trees along its bank. In mountainous areas, streams may only tumble over and around rock with very little vegetation. Some streams actually dry up during periods of low rainfall. These are called *intermittent* streams.

As streams increase in flow and join with other streams, a branching network is established, much like the branches of a tree. This network from headwater streams to the river mouth is called a river system. A numbering method has been devised to describe the relative position of a stream within a river system. The beginning of a stream is described as a *first-order* stream. Two first-order streams join to form a *second-order* stream, and two second-order streams become a *third-order* stream, and so on. The Mississippi River near its mouth is considered a *twelfth-order* river!

Rivers are inseparable from the land through which they flow.

The land area that drains rain and snowmelt to a river is called a watershed. Within a watershed each tributary is part of a smaller watershed. Every person on earth lives within a watershed. Do you know what your watershed looks like? What are some primary uses of water in the headwaters of your watershed? Near the mouth? How does land use affect your river? (See Chapter 7 for more information about land uses and their impact on rivers.)

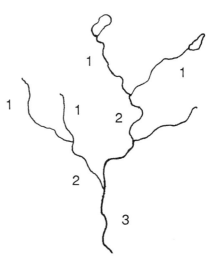

Figure 2.1. Stream ordering in the headwaters of a river.

Rivers are some of the oldest natural features on Earth.

Rivers stretch back in time millions of years. Just imagine the age of the Colorado River, as it has cut a channel almost a mile deep called the Grand Canyon. Because rivers around the world are very old, they have provided an environment for the evolution of very specific organisms that can live underwater. In Chapter 6 you will learn more about aquatic insects and other organisms that have adapted to living in flowing water. It is interesting to observe that many types of aquatic insects are similar in appearance (such as mayflies) whether they come from a river in Africa, the Middle East, Europe, Asia, Australia, Latin America, or North America.

Rivers were also the sites of early civilizations because they provided freshwater for drinking, as well as fish and other food. Floodplains along the river provided rich areas to grow crops, and rivers themselves were used for transportation.

Because people around the world have lived near rivers for centuries, they play an important role in many of the world's cultures. In India, people believe that the Ganges River is a holy river and that the river has the power to wash away sin. Before pollution and dams, natives in North America celebrated the return of salmon every year to the rivers where they were born. For many tribes, salmon was a major source of food. (Please refer to Chapter 9 for more background on the importance of rivers around the world.)

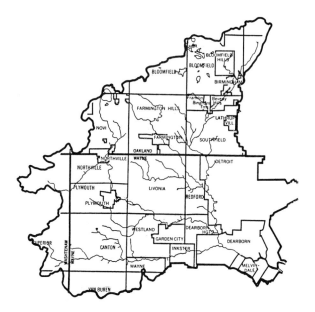

Figure 2.2. The upper reaches of a watershed showing the network of branching streams that join to form the Rouge River in Michigan.

Rivers are some of the most used and abused natural features on Earth.

Many watersheds have been altered as a result of human needs for water, food, recreation, transportation, manufactured goods, and other amenities. These growing demands have led to pollution of streams and rivers, and unwise land uses that further degrade water quality.

Physical, chemical, and biological measurements are the tools necessary to recognize trends in water quality. It is also important to learn how natural river ecosystems work in order to understand the changes that most of our rivers have undergone. Finally, it is important to recognize how our daily activities may affect water quality, so that we can take action in our own lives to restore the health of our rivers.

To integrate some of the factors responsible for river water quality, this manual considers:

➤ nine physical-chemical measures that form a Water Quality Index;

➤ aquatic insects as representative of the aquatic community;

➤ land uses that influence the nature of rivers;

Figure 2.3. High school students monitoring their community's windsurfing area to determine if the water quality is safe for body contact recreational activities.

➤ human impacts on the stream and river environment;

➤ stream surveys to determine changes in the land bordering rivers;

➤ the international dimension of rivers; and

➤ ways to study rivers to enhance educational programs.

How to Measure Water Quality Using a Water Quality Index

Many factors can affect the water quality of a river system. The conditions of a river can fluctuate periodically, so we must measure water quality periodically to look for trends. Water that is determined to be safe for one use may be unacceptable for another purpose. In fact, many water quality experts refer to their measurements in terms of a specific use.

The categories used for making recommendations on water uses are: water supply for domestic and industrial use; recreation for total body contact like swimming, water skiing, skin diving, and windsurfing; partial body contact recreation, like fishing, and boating; protection of aquatic organisms such as fish; agricultural uses like livestock watering, irrigation, and spraying; and, commercial uses such as navigation, hydroelectric, and steam-generated electric power.

In an attempt to devise a system to compare rivers in various parts of the country, the National Sanitation Foundation (NSF) created and designed a standard index, called the Water Quality Index (WQI). The WQI is one of the most widely used of all existing water quality indices. It was developed in 1970, and can be used to measure water quality changes in a particular river reach over time, compare water quality from different reaches of the same river, and even compare water quality of different rivers. The results can also be used to determine if a particular stretch of the river is healthy.

To determine the WQI, nine tests are performed. These include dissolved oxygen, fecal coliform, pH, biochemical oxygen demand (5-day), temperature, total phosphate, nitrates, turbidity, and total solids. After completing the nine tests, the results are recorded and transferred to a weighting curve chart where a numerical value is obtained. For each test, the numerical value or Q-value is multiplied by a "weighting factor." For example, dissolved oxygen has a relatively high weighting factor (.17) and therefore is more significant in determining water quality than the other tests. The nine resulting values are then added to arrive at an overall water quality index (WQI).

Test Results	Raw Data (Column A)	Q-Value (Column B)	Weighting Factor (Column C)	TOTAL (Column D)
1. Dissolved Oxygen	60% sat.	58	0.17	9.86
2. Fecal Coliform	20 colonies/100 ml	62	0.16	9.92 (est)
3. pH	8 units	85	0.11	9.35
4. B.O.D.	6 mg/L	51	0.11	5.61
5. Temperature	0°C ΔT	92	0.11	10.12
6. Total Phosphate	.4 mg/L	70	0.10	7.00
7. Nitrates	.8 mg/L	96	0.10	9.60
8. Turbidity	3/feet (100/cm)	76	0.08	6.08
9. Total Solids	709 mg/L	20	0.07	1.40

Overall Water Quality Index   70.26
(est.)

Figure 2.4. Water quality data collected by high school students at a popular water sports area where concessionaires offered windsurfing rentals and lessons.

If you are unable to run all 9 water quality tests and you want to estimate the "Overall Water Quality Index" follow this example. Let's say that the test results on page 10 were complete except there was no fecal coliform data. Simply add the totals from column D for the other eight indicators—that would be 59.02 or 59. The total of the nine weighting factors = 1.0. If fecal coliform is missing (it's weighting factor is .16) the total of the remaining weighting factors would be .84. To factor in the missing data, multiply the total derived from the eight indicators or 59 x 1.19 (the inverse of .84) = 70.21 or 70.

As you run the water quality tests, use the weighted curve charts, and derive a water quality index, this entire process will become clear. The instructions given in this manual to run the tests for dissolved oxygen, biochemical oxygen demand, pH, nitrates, and total phosphate are based upon LaMotte kits. LaMotte is not the only company, however, from which to purchase water quality tests. Additional companies whose equipment is quite compatible are listed in Appendix A.

The nitrate and total phosphate tests require sensitive testing procedures to obtain accurate results because these nutrients are often found in very low concentrations. Portable nitrate and total phosphate kits are much simpler to use but are not as accurate as the tests described in *Standard Methods for the Examination of Water and Wastewater.* (The local wastewater treatment plant will have a copy of this manual.) Field equipment such as LaMotte kits, however, are very convenient and provide accurate results for the other parameters.

Origins of the WQI

To develop the WQI, the National Sanitation Foundation selected 142 people who represented a wide range of positions at the local, state, and national level. Through a series of questionnaires, each panelist was asked to consider 35 water quality tests for possible inclusion in an index. This number was finally reduced to the nine tests stated above.

The weighting curve charts are also the product of repeated questionnaires. Respondents were asked to graph the level of water quality ranging from "0" (worst) to "100" (best) as determined by the raw test data. The weighted curves are a result of averaging all of the curves drawn. These curves reflect the best professional judgement of the respondents to an arbitrary scale of water quality (0–100).

Water Quality Index Ranges

90–100	Excellent
70– 90	Good
50– 70	Medium
25– 50	Bad
0– 25	Very Bad

It should be noted that field test kits are quite accurate for testing the quality of fresh, brackish, and salt water.

Sampling

It is important to exercise care in the way samples are collected for analysis. A collected sample should be representative of the river reach being tested. Analytical values derived from river samples may vary with depth, velocity of current, and the distance the sample was taken from shore.

Figure 2.5. A homemade device for collecting water samples can be fashioned using an extension arm and rubber tubing for holding the sample bottle.

With these thoughts in mind, sampling from shore is not the preferred method of sampling. Near-shore samples may not be representative of the river at that location. If possible, water samples should be collected from a bridge spanning the river, from a boat, or off the end of a dock. A rule of thumb for sampling is to sample midway across the river and below the surface.

A simple sampling device can be constructed from a series of metal rods that can be extended and rubber tubing attached that holds the sample bottle (see figure 2.5). This device might be extended out from shore if no bridges are available and particularly if the river is narrow or shallow. A golf ball retriever can also be adapted very easily for this purpose.

Figure 2.6. A Secchi disk for measuring turbidity designed and made in the school's metal and wood shop by a high school student.

Building sampling devices are worthwhile projects that students in metal shop or related courses might want to undertake. If not, local hardware stores can often construct these devices given the measurements.

The samples most often taken by students are called *grab samples,* which are single samples representative of the river at a particular time and place. To improve the reliability of the water quality data collected, *replicate samples* (additional samples) should be taken from a particular location, at the same time. For example, more than one group could run a dissolved oxygen test at the same river location.

A mixture of grab samples collected at the same sampling point at different times is called a *composite sample.* If a number of classes collect grab samples over the course of a day at the same sampling location, then this would also be a form of composite sampling, but only if an average value is calculated from all of the samples. For example, if four different classes each ran a nitrate test at the same sampling point and then combined these individual nitrate values to compute an average value, then this would be a form of composite sampling.

It is important to remember that the shorter the time between sampling and analysis, the more reliable the results will be. The dissolved oxygen, temperature, and pH tests should be completed at the sampling site, because the results of these tests change quickly when water samples sit after being collected. Specific sampling considerations will be covered under the appropriate water quality tests.

If the purpose of sampling is to note change in the Water Quality Index from year to year, then it is important to run the water quality tests at approximately the same week and time of the day each year.

If you suspect pollution entering the river from a pipe, then it is a good strategy to sample just above the pipe, immediately downriver from the pipe, and much further downriver (but above other suspected point sources). You may decide to monitor for one or two water quality tests rather than the entire WQI. By monitoring upriver, and at points downriver from a suspected pollution source and repeating tests, it is possible to verify such a source.

Safety Guidelines

We strongly urge you to review the following safety guidelines for sampling and running water quality tests before you begin a monitoring program. The following are general guidelines; more specific safety directives are included under the appropriate water quality tests.

Safety in Sampling

1. Whenever sampling, all skin that could potentially be in contact with the water should be covered. This means wearing rubber gloves and boots if necessary. The use of an extension device for sampling will minimize exposure to the river.

2. When monitoring close to a wastewater treatment plant, and particularly when a strong wind is blowing from the direction of the plant, surgical masks should be considered to protect against aerosols. (Aerosols are windborne contaminants that can be breathed deeply into the lungs if present.)

3. Contact your nearest Health Department or Department of Natural Resources, or the U.S. Environmental Protection Agency (EPA) for specific warnings about local rivers. Some stretches of river and land bordering rivers may contain dangerous levels of toxic contaminants in the sediment. If in doubt, please consult the local authorities.

4. Avoid sampling from heavily used bridges, and only do so after consulting the local public works department. Sampling sites with steep banks should also be avoided if possible.

Safety in Running the Tests

1. There should be some Material Safety Data Sheets (MSDS) packed in with each water quality test kit. These sheets provide very specific first aid and chemical information if someone ingests one of the chemicals, or if it comes in contact with someone's eyes or skin. Such sheets should be reproduced and displayed for others to see.

2. Ensure that students and others understand from the beginning the danger of treating these chemicals casually or endangering others through "horseplay."

3. Safety goggles should be worn, particularly when running water quality tests that require shaking or swirling a chemical mixture (dissolved oxygen, biochemical oxygen demand, pH, nitrates, and total phosphate.)

4. Wash your hands after running every water quality test. Avoid placing hands in contact with eyes or mouth during monitoring.

Figure 2.7. Students using safety goggles while running the dissolved oxygen test.

5. Follow the general safety guidelines for your particular school.

6. Dispose of spent chemicals in an environmentally sound manner. (See specific instructions for individual tests.)

7. In general, the following items will help to ensure a safe monitoring experience:

   a. Safety goggles for each student

   b. Clean pail or bucket for washing hands

   c. Jug of clean water for washing hands

   d. Soap (biodegradable if possible)

   e. Towels

   f. Waste container for liquid chemical waste (except nitrate waste; see item "j" below)

   g. Plastic gloves

   h. Eye wash bottle

   i. First aid kit

   j. Hazardous waste container clearly marked (from nitrate test liquid waste) and then deposited in accordance with hazardous waste guidelines.

This manual is the product of years of experience working in both public and private school settings. Sampling procedures, safety precautions, and testing procedures contained in the manual reflect the very real responsibilities, and economic and logistical concerns present in diverse schools.

For further information about any particular test, please consult *Standard Methods for the Examination of Water and Wastewater* (16th ed.), American Public Health Association, Inc., New York.

Nine Water Quality Tests— What They Mean, and How to Do Them

Dissolved Oxygen

Dissolved oxygen (DO) is essential for the maintenance of healthy lakes and rivers. The presence of oxygen in water is a positive sign, the absence of oxygen a signal of severe pollution. Rivers range from high to very low levels of dissolved oxygen—so low, in some cases, that they are practically devoid of aquatic life.

Most aquatic plants and animals need oxygen to survive. Fish and some aquatic insects have gills to extract life-giving oxygen from the water. Some aquatic organisms, like pike and trout, require medium-to-high levels of dissolved oxygen to live. Other animals, like carp and catfish, flourish in waters of low dissolved oxygen. Waters of consistently high dissolved oxygen are usually considered healthy and stable ecosystems capable of supporting many different kinds of aquatic organisms.

Sources of Dissolved Oxygen

Much of the dissolved oxygen in water comes from the atmosphere. Waves on lakes and slow-moving rivers, and tumbling water on fast-moving rivers act to mix atmospheric oxygen with water. Algae and rooted aquatic plants also deliver oxygen to water through photosynthesis.

In general, rooted aquatic plants are more abundant in lakes and impounded rivers than in rivers with significant current. Large daily fluctuations in dissolved oxygen are characteristic of bodies of water with extensive plant growth. Dissolved oxygen levels rise from morning through the afternoon as a result of photosynthesis, reaching a peak in late afternoon. Photosynthesis stops at night, but plants and animals continue to respire and consume oxygen, As a result, dissolved oxygen levels fall to a low point just before dawn. Dissolved oxygen levels may dip below 4 mg/liter in such waters—the minimum amount needed to sustain warmwater fish like bluegill, bass, and pike.

Physical Influences on Dissolved Oxygen

Water temperature and the volume of water moving down a river (discharge) affect dissolved oxygen levels. Gases, like oxygen, dissolve more easily in cooler water than in warmer water. In temperate areas, rivers respond to changes in air temperature by cooling or warming.

DO

River discharge is related to the climate of an area. During dry periods, flow may be severely reduced, and air and water temperatures are often higher. Both of these factors tend to reduce dissolved oxygen levels. Wet weather or melting snows increase flow, with a resulting greater mixing of atmospheric oxygen.

Human-Caused Changes in Dissolved Oxygen

The main factor contributing to changes in dissolved oxygen levels is the build-up of organic wastes. Organic wastes consist of anything that was once part of a living plant or animal, including food, leaves, feces, etc. Organic waste can enter rivers in many ways, such as in sewage, urban and agricultural runoff, or in the discharge of food processing plants, meat packing houses, dairies, and other industrial sources.

A significant ingredient in urban and agricultural runoff are fertilizers that stimulate the growth of algae and other aquatic plants. (See the sections in this chapter on biochemical oxygen demand, nitrate and total phosphate for more information.) As plants die, aerobic bacteria consume oxygen in the process of decomposition. Many kinds of bacteria also consume oxygen while decomposing sewage and other organic material in the river.

Changes in Aquatic Life

Depletions in dissolved oxygen can cause major shifts in the kinds of aquatic organisms found in water bodies. Species that cannot tolerate low levels of dissolved oxygen—mayfly nymphs, stonefly nymphs, caddisfly larvae, and beetle larvae—will be replaced by a few kinds of pollution-tolerant organisms, such as worms and fly larvae. (See Chapter 6 for more information on these and other aquatic organisms.) Nuisance algae and anaerobic organisms (that live without oxygen) may also become abundant in waters with low levels of dissolved oxygen.

Calculating Percent Saturation

The percent saturation of water with dissolved oxygen at a given temperature is determined by pairing temperature of the water with the dissolved oxygen value, after first correcting your dissolved oxygen measurement for the effects of atmospheric pressure. This is done with the use of the correction table in Figure 3.1, and the percent saturation chart in Figure 3.2.

DO

Rivers that consistently have a dissolved oxygen value of 90 percent or higher are considered healthy, unless the waters are supersaturated due to cultural eutrophication. Rivers below 90 percent saturation may have large amounts of oxygen-demanding materials, i.e. organic wastes.

To calculate percent saturation, first correct your dissolved oxygen value (milligrams of oxygen per liter) for atmospheric pressure. Turn to Figure 3.1. Using either your atmospheric pressure (as read from a barometer) or your local altitude (if a barometer is not available), read across to the right hand column to find the correction factor. Multiply your dissolved oxygen measurement by this factor to obtain a corrected value.

Now turn to the chart in Figure 3.2. Draw a straight line between the water temperature at the test site and the corrected dissolved oxygen measurement, and read the saturation percentage at the intercept on the sloping scale.

Example:

Let's say that your dissolved oxygen value was 10 mg/L, the measured water temperature was 15°C, and the atmospheric pressure at the time of sampling was 608 mmHg. From the table in Figure 3.1, the correction factor is 80 percent, which multiplied by 10 mg/L gives a corrected dissolved oxygen value of 8 mg/L. Drawing a straight line between this value and 15°C gives a percent saturation of about 80 percent.

How might you interpret these results? At the relatively cool temperature of 15°C, one would expect a river to have a dissolved oxygen value higher than 80 percent. It would appear that something is using up oxygen in the water.

Atmospheric Pressure (mmHg)	Equivalent Altitude (ft.)	Correction Factor
775	-540	1.02
760	0	1.00
745	542	.98
730	1094	.96
714	1688	.94
699	2274	.92
684	2864	.90
669	3466	.88
654	4082	.86
638	4756	.84
623	5403	.82
608	6065	.80
593	6744	.78
578	7440	.76
562	8204	.74
547	8939	.72
532	9694	.70
517	10,472	.68

DO

Figure 3.1. Correction table for dissolved oxygen measurements.

Sampling Procedures

Because DO levels vary so much according to time, weather, and temperature, this test should be run during the same period (week and time of day) if yearly comparisons are to be made. In rivers, there is usually adequate mixing of water from the surface to the river bottom. However, in impounded river reaches or in very large deep rivers there may be little mixing of the water. This could cause differences in DO measurements from the surface to the river bottom.

It is best to sample away from shore and below the water surface. In free-flowing rivers with good mixing, samples taken beneath the surface and in the current will probably be representative samples. In slow-moving river reaches and in impounded river areas with little mixing, it is very important to sample away from shore and to sample at various depths. Shore sampling will probably not provide a representative sample in these waters. Nor will a sample taken from only one depth, since aquatic vegetation produces oxygen near the surface, while decaying vegetation on the bottom consumes oxygen through the respiration of aerobic (oxygen-dependent) bacteria.

The extended rod sampler with an elastic strap or wire basket can be used to hold a dissolved oxygen bottle (see Figure 2.5). If no bridge is available to sample from, or if the bridge carries too much traffic to be used safely, or is too high above the water, perhaps a boat may be found. If neither a bridge nor a boat are available, the best option is to extend the rod sampler from shore as far as possible and take a sample beneath the river surface. A dissolved oxygen sample can be obtained near shore without a sampling device, but keep in mind that it is probably not a representative sample.

DO

Remember that the dissolved oxygen test should be run immediately after sampling.

> **Warning:**
>
> *Please wear protective gloves. If your skin comes into contact with any reagent, rinse this area liberally with water. First aid directions are included on the reagent containers. SAFETY GOGGLES SHOULD BE WORN WHILE SHAKING THE DISSOLVED OXYGEN BOTTLE.*

Dissolved Oxygen Testing Procedure

1. If you have a barometer, record the atmospheric pressure. Remove the cap and immerse the DO bottle beneath the river's surface. Use gloves to avoid contact with the river.

2. Allow the water to overflow for two to three minutes. (This will ensure the elimination of air bubbles.)

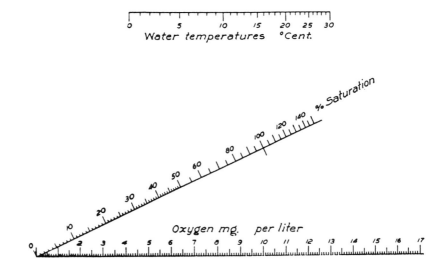

Figure 3.2. Level of oxygen saturation chart.

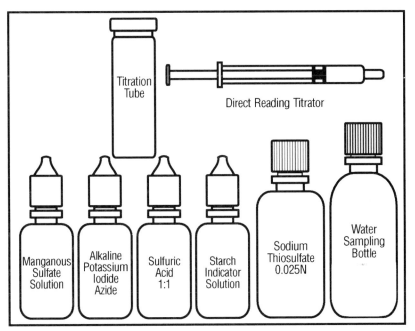

Titration Tube

Direct Reading Titrator

DO

Water Sampling Bottle

Sodium Thiosulfate 0.025N

Manganous Sulfate Solution

Alkaline Potassium Iodide Azide

Sulfuric Acid 1:1

Starch Indicator Solution

Figure 3.3. Dissolved oxygen test kit items.

3. Make sure no air bubbles are present when you take the bottle from the river.

4. Add 8 drops of Manganous Sulfate Solution and 8 drops of Alkaline Potassium Iodide Azide.

5. Cap the bottle, making sure no air is trapped inside, and invert repeatedly to fully mix. Be very careful not to splash the chemical-laden water. *Wash your hands if you contact this water.* If oxygen is present in the sample, a brownish-orange precipitate will form (floc). The first two reagents "fix" the available oxygen.

6. Allow the sample to stand until the precipitate settles halfway. When the top half of the sample turns clear, shake again, and wait for the same changes.

7. Add 8 drops of Sulfuric Acid 1:1 Reagent. Cap and invert repeatedly until the reagent and the precipitate have dissolved. A clear yellow to brown-orange color will develop depending on the oxygen content of the sample.

DO

> **Note:**
>
> Following completion of step 7, contact between the water sample and the atmosphere will not affect the test result. Once the sample has been "fixed" in this manner, it is not necessary to perform the actual test procedure immediately. Several samples can be collected and "fixed" in the field, and then carried back to a testing station or laboratory where the titration procedure is to be performed.

8. Fill the titration tube to the 20 mL line with the "fixed" sample and cap.

9. Fill the Direct Reading Titrator with Sodium Thiosulfate 0.025N Reagent. Insert the Titrator into the center hole of the titration tube cap. While gently swirling the tube, slowly press the plunger to titrate until the yellow-brown color is reduced to a very faint yellow.

> Note: if the color of the fixed sample is already a very faint yellow, skip to step 10.

10. Remove the cap and Titrator. Be careful not to disturb the Titrator plunger, as the titration begun in step 8 will continue in step 11. Add 8 drops of Starch Indicator Solution. The sample should turn blue.

11. Replace the cap and Titrator. Continue titrating until the sample changes from blue to a colorless solution. Read the test result where the plunger tip meets the scale. Record as mg/L (ppm) dissolved oxygen.

> Note: Each minor division on the Titrator scale equals 0.2 mg/L (0.2 ppm).

> Note: If the plunger tip reaches the bottom line on the Titrator scale (10) before the endpoint color change occurs, refill the Titrator and continue the titration. When recording the test result, be sure to include the value of the original amount of titrant dispensed.

FC

Figure 3.4. Check each other to make certain that all water quality tests are performed accurately and safely.

Fecal Coliform

Fecal coliform bacteria are found in the feces of humans and other warm-blooded animals. These bacteria can enter rivers through direct discharge from mammals and birds, from agricultural and storm runoff carrying wastes from birds and mammals, and from human sewage discharged into the water.

Fecal coliform by themselves are not usually pathogenic. Pathogenic organisms include bacteria, viruses, and parasites that cause diseases and illnesses. Fecal coliform bacteria naturally occur in the human digestive tract, and aid in the digestion of food. In infected individuals, pathogenic organisms are found along with fecal coliform bacteria.

If fecal coliform counts are high (over 200 colonies/100 mL of water sample) in the river, there is a greater chance that pathogenic organisms are also present. A person swimming in such waters has a greater chance of getting sick from swallowing disease-causing organisms, or from pathogens entering the body through cuts in the skin, the nose, mouth, or the ears. Diseases and illness such as typhoid fever, hepatitis, gastroenteritis, dysentery, and ear infections can be contracted in waters with high fecal coliform counts.

Pathogens are relatively scarce in water, making them difficult and time-consuming to monitor directly. Instead, fecal coliform levels are monitored, because of the correlation between fecal coliform counts and the probability of contracting a disease from the water.

Cities and suburbs sometimes contribute human wastes to local rivers through their sewer systems. A sewer system is a network of underground pipes that carry wastewater.

In a *separate sewer system,* sanitary wastes (from toilets, washers, and sinks) flow through sanitary sewers and are treated at the wastewater treatment plant. Storm sewers carry rain and snowmelt from streets, and discharge untreated water directly into rivers. Heavy rains and melting snow wash bird and pet wastes from sidewalks and streets, and may "flush out" fecal coliform from illegal sanitary sewer connections into the storm sewers.

In a *combined sewer system,* sanitary wastes *and* storm runoff are treated at a wastewater treatment plant. After a heavy rain, untreated or inadequately treated waste may be diverted into the river to avoid flooding the wastewater treatment plant. To avoid this problem, some cities have built retention basins to hold excess wastewater and prevent untreated wastes from being discharged into rivers. Without retention basins, heavy rain conditions can result in high fecal coliform counts downstream from sewage discharge points. That is why it is important to note weather conditions on the days before a fecal coliform measurement.

Coliform Standards (in colonies/100 ml)

Drinking water . . . . 1 TC
Total body contact (swimming) . . . . 200 FC
Partial body contact (boating) . . . . 1000 FC
Treated sewage effluent . . . . Not to exceed 200 FC

*Total coliform (TC) includes bacteria from cold-blooded animals and various soil organisms. According to recent literature, total coliform counts are normally about 10 times higher than fecal coliform (FC) counts.

Figure 3.5. Fecal and total coliform standards for water used for drinking and recreation, and treated sewage.

Sampling Procedures

1. Remove the stopper or cap just before sampling and avoid touching the inside of the cap.

2. If sampling by hand, use gloves and hold the bottle near its base. Plunge it (opening downward) below the water surface, then turn the bottle underwater into the current and away from you.

3. Avoid sampling the water surface because the surface film often contains greater numbers of fecal coliform bacteria than is representative of the river.

4. Also, avoid sampling the sediments for the same reason, unless this is intended. The same general sampling procedures apply when using the extended rod sampler.

5. When collecting samples, leave some space in the sample container (an inch or so) to allow mixing of the sample before pipetting.

FC

It is a good idea to collect several samples from any single location on the river to minimize the variability that comes with sampling for bacteria. If possible, sterilization should occur between sampling sites. *Ideally, all samples should be tested within one hour of collection. If this is not possible, the sample bottles should be placed in ice and tested within six hours.*

Fecal Coliform Testing Procedure

1. First, sanitize the forceps by dipping forceps in alcohol, then burning alcohol off with a flame (an alcohol lamp works well). Do not place the hot forceps back into the alcohol.

2. Using the sanitized forceps, place an absorbent pad in the presterilized petri dish. Be careful not to touch the pad with your fingers.

3. Unscrew the neck of the broth plastic tube (ampoule) or use an ampoule breaker if needed, and drain the broth onto the pad. (The broth is liquid food for fecal coliform bacteria.) Put the top on the petri dish and set aside.

4. Sanitize forceps with alcohol and flame again.

5. Unscrew the top half of the filtration system and place a sterile filter paper on top of the filtration system's membrane with forceps, grid side up. Be sure the filter lies completely flat with no wrinkles.

A word about sterilization . . .

It is essential to sterilize sample bottles, pipettes, and filtration system before sampling. Sterilization can be accomplished by using an autoclave, 121° C for 15 minutes. If an autoclave is not available, the home economics department may have a pressure cooker that they might be willing to lend to the water quality monitoring program. If a pressure cooker is used, please be sure that it has a working pressure gauge. The gauge may be checked with the county cooperative extension service. The pressure cooker should be run at 15 psi. to properly sterilize sample bottles, pipettes and filtration system.

If these two pieces of equipment are unavailable, an oven can be used. The oven must attain a temperature of 170° (±10°C) for not less then 60 minutes. The plastic filtration system cannot, however, be placed in a dry oven because the system will melt. The same holds true for plastic sampling bottles. The filtration system can, however, be placed in boiling water for 5 minutes to sanitize it. Petri dishes, culture media, absorbent pads, and filters are presterilized and packaged. Equipment that has been inadequately sterilized may interfere with fecal coliform growth.

. . . and sampling design

If the purpose of sampling is to determine fecal coliform levels at a river reach, then samples should be taken beneath the water surface and in the current (if there is one). If the purpose of sampling is to confirm suspected sources of fecal coliform contamination, then samples should be taken just downriver from the source (like the mouth of a storm drain), and other samples should be taken upriver from the source for comparison.

There is also wet-weather sampling and dry-weather sampling. Wet-weather sampling involves sampling during and just after a rainstorm, often at timed intervals. It is done if fecal coliform contamination is suspected from storm drains carrying urban stormwater runoff. Wet-weather samples can then be compared to samples taken during a period of dry weather (dry-weather samples). The bottles used for the dissolved oxygen test might also be used for the fecal coliform test.

Try to avoid sampling stagnant areas of rivers. The extended rod sampler (Figure 2.5) is an effective device for obtaining a sample in the current. If sampling rivers in which little current exists, push the sample bottle underwater away from your body, thereby creating a current.

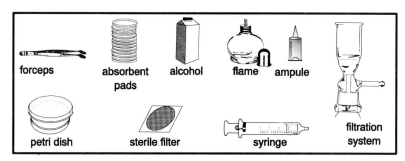

Figure 3.6. Fecal coliform test equipment items.

FC

6. Screw on the top half of the filtration system to the bottom half.

7. Before taking a sample, use a pipette to rinse the filtration system with a small amount of distilled water. Add the water through the hole in the top of system. (There should be two or three rubber stoppers on top of the filtration system, and one hole without a stopper. See Figure 3.8.)

8. Determine the desired volume of water (in mL) to be tested based upon the water source. (See Figure 3.7 for suggested sample volumes sizes.) Place the pointed end of the pipette into the water to be sampled and lower into the water until the desired sample size, as shown by volume markings on the side of pipette, has been drawn into the pipette. A rubber bulb attached to the top of the pipette may be required to obtain the desired volume. When there are high numbers of fecal coliform, a proper sample should not exceed 60 colonies on the petri plate. The higher range affects the colony sheen or color development resulting in errors in making a proper count.

Drinking water	100.0					
River water	5.0	2.0	1.0	0.1		
Stormwater runoff			1.0	0.1	0.01	
Raw Sewage, CSO's			1.0	0.1	0.01	0.001

Adapted from *Standard Methods for the Examination of Water and Wastewater.*

Figure 3.7. Suggested fecal coliform sample volumes in mL.

Figure 3.8. Putting sample of water into filtration system.

9. Place the end of the pipette into the open hole on top of the filtration system, and release the water sample into the funnel. See Figure 3.8.

10. With the filtration system level, use the suction pump and draw all of the sample and distilled water through the filter while swirling so that the number of bacteria adhering to the upper filtration system is reduced. (Warning: be careful when pushing the plunger back into the syringe; you want to avoid pushing air back into the filtration unit and forcing the filter off the membrane.) Draw the water through the filter until it appears dry.

11. Unscrew the top half of the funnel, and carefully remove the filter with the sanitized forceps.

If you do not have a water bath, one might be available at a local university, community sewage treatment plant, or local laboratory. If no water bath is available in your community you might try a hot air incubator if it holds temperature. Recognizing that a water bath is relatively expensive, multiple schools within a watershed involved in a water monitoring program might want to purchase one water bath for the entire program.

FC

12. Open the top of the petri dish, and slide the filter across and into the dish, with the grid side up. Petri dishes should be incubated within 30 minutes of filtering the sample; this will ensure heat shock of any non-fecal coliform organisms. *Be sure to record the date, site, and volume of sample on the frosted part of the petri dish.*

13. Enclose the petri dish in a waterproof bag (to avoid leakage) and then put into the water bath. Dishes may also be sealed with waterproof tape (freezer tape) to avoid leakage. Incubate for 24 hours (±2 hours) at 44.5°C. (Temperature must be maintained within a range of ±0.25°C of 44.5°C.) Petri dishes should be inverted during incubation to avoid condensation. *Please wash your hands after this test.*

14. After incubation, carefully count the bacterial colonies on the filter, using a magnifying glass (10x) or unaided eye (see Figure 3.9). You might want several people to verify the bacterial count. Each bluish spot is counted as one fecal coliform

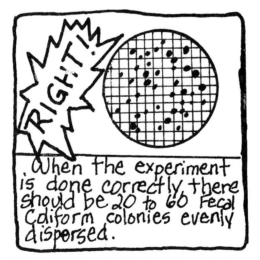

Figure 3.9. Counting fecal coliform colonies.

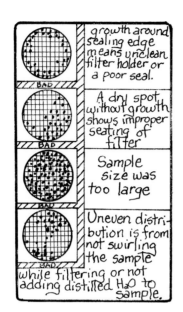

growth around sealing edge means unclean filter holder or a poor seal.

A dry spot without growth shows improper seating of filter

Sample size was too large

Uneven distribution is from not swirling the sample while filtering or not adding distilled H₂O to sample.

Figure 3.10. Troubleshooting fecal coliform cultures.

colony. Cream or gray-colored colonies are nonfecal coliform. Fecal coliform colonies should be examined within 20 minutes to avoid color changes that occur with time. Some common fecal coliform culture problems can be seen in Figure 3.10.

15. *It is important to report the highest fecal coliform value rather than an average value.*

Detection of Waterborne Coliform and Fecal Coliforms with Coliscan Easy Gel

Available through Earth Force,
1908 Mount Vernon Ave.
2nd Floor
Alexandria, VA
Tel: 703-299-9400
FAX: 703-299-9485

The *Volunteer Monitor* recently brought to our attention a new process for coliform and fecal coliform testing that does not require an incubator or water bath.

1. Use a sterile calibrated dropper to collect a 1 mL water sample and deposit the sample into a bottle containing liquid coliscan medium (this procedure may be done in the field and the coliscan-water mix can be kept on ice until it is returned to the lab).

2. Pour the coliscan water mix into a pre-treated petri dish and swirl to cover entire bottom of petri dish.

3. Place the petri dish containing the coliscan-water mix in a warm place and incubate for 24–48 hours (this is best done in a place such as an incubator which holds the temperature in a range of 85°–95° F).

4. Count the red colonies in the petri dish as coliforms and the purple colonies as fecal coliforms (E. coli). (White or blue-green colonies should be noted, but they are not classified as coliforms or fecal coliforms.)

pH

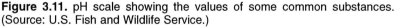

Water (H_2O) contains both H^+ (hydrogen) ions and OH^- (hydroxyl) ions. The pH test measures the H^+ ion concentration of liquids and substances. Each measured liquid or substance is given a pH value on a scale that ranges from 0 to 14.

Pure deionized water contains equal numbers of H^+ and OH^- ions, and has a pH of 7. It is considered neutral, neither acidic or basic. If a water sample has more H^+ than OH^- ions, it is considered acidic and has a pH less than 7. If the sample contains more OH^- ions than H^+ ions, it is considered basic with a pH greater than 7. (See pH scale in Figure 3.11.)

It is important to remember that for every one unit change on the pH scale, there is approximately a ten-fold change in how acidic or basic the sample is. For example, the average pH of rainfall over much of the northeastern United States is 4.3, or roughly ten times more acidic than normal rainfall of 5.0-5.6. Lakes of pH 4 (acidic) are roughly 100 times more acidic than lakes of pH 6.

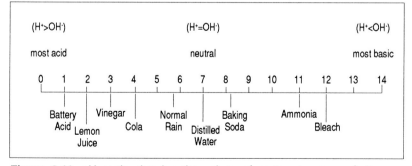

Figure 3.11. pH scale showing the values of some common substances. (Source: U.S. Fish and Wildlife Service.)

Human-Caused Changes in pH

In the U.S., the pH of natural water is usually between 6.5 and 8.5, although wide variations can occur. Increased amounts of nitrogen oxides (NO_x^-) and sulfur dioxide (SO_2^-), primarily from automobile and coal-fired power plant emissions, are converted to nitric acid and sulfuric acid in the atmosphere. These acids combine with moisture in the atmosphere and fall to earth as acid rain or acid snow.

Acid rain is responsible for thousands of lakes in eastern Canada, northeastern United States, Sweden, and Finland becoming acidic. In many areas of the United States, the type of rocks and minerals present determine the acidity of the local water. If limestone is present, the alkaline (basic) limestone neutralizes the effect the acids might have on lakes and streams.

pH

The areas hardest hit by acid rain and snow are downwind of urban/industrial areas and do not have any limestone to reduce the acidity of the water.

Changes in Aquatic Life

Changes in the pH value of water are important to many organisms. Most organisms have adapted to life in water of a specific pH and may die if it changes even slightly. This has happened to brook trout in some streams in the Northeast.

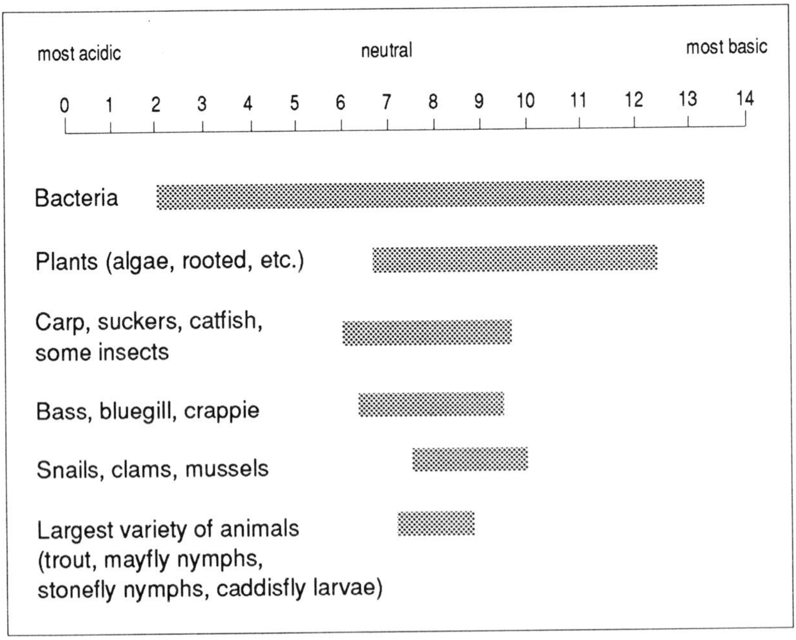

Figure 3.12. pH ranges that support aquatic life.

At extremely high or low pH values (e.g., 9.6 or 4.5) the water becomes unsuitable for most organisms (see Figure 3.12). Serious problems occur in lakes with a pH below 5, and in streams that receive a massive acid dose as the acidic snow melts in the spring. Immature stages of aquatic insects and young fish are extremely sensitive to pH values below 5.

Very acidic waters can also cause heavy metals, such as copper and aluminum, to be released into the water. Heavy metals can accumulate on the gills of fish or cause deformities in young fish, reducing their chance of survival. (See Chapter 5 for more information about toxics.)

pH Sampling Procedure

Like the sample collected for the dissolved oxygen test, the water sample for the pH test should be collected away from the river bank and below the surface. If possible, use an extension rod sampler like the one shown in Figure 2.5.

The sample must be measured immediately because changes in temperature can affect the pH value. If pH must be measured later, the sample should be placed on ice, and measured as soon as possible.

If the pH of the same water sample is tested more than once, the most common pH value (the mode) should be reported, not the average value.

pH Testing Procedure

1. Rinse each test tube with the water sample. Gloves should be worn to avoid skin contact with the water.
2. Fill the tube to the 5 mL line with sample water.
3. While holding dropper bottle vertically, add 10 drops of Wide Range Indicator Solution.
4. Cap and invert several times to mix.
5. Insert the tube into the Wide Range pH Comparator. Hold the comparator up to a light source. Match the sample color to a color standard.
6. Record the pH value.
7. Wash your hands.

I **Figure 3.13.** Students recording their pH readings.

Biochemical Oxygen Demand (BOD 5-Day)

When organic matter decomposes, it is fed upon by aerobic bacteria. In this process, organic matter is broken down and oxidized (combined with oxygen). Biochemical oxygen demand is a measure of the quantity of oxygen used by these microorganisms in the aerobic oxidation of organic matter.

When aquatic plants die, they are fed upon by aerobic bacteria. The input of nutrients into a river, such as nitrates and phosphates, stimulates plant growth. Eventually, more plant growth leads to more plant decay. Nutrients, then, can be a prime contributor to high biochemical oxygen demand in rivers.

Impounded river reaches also collect organic wastes from upriver that settle in quieter water. The bacteria that feed on this organic waste consume oxygen. Percent saturation (dissolved oxygen) values in waters with much plant growth and decay often fall below 90 percent.

Sources of Organic Matter

There are natural sources of organic material which include organic matter entering lakes and rivers from swamps, bogs, and vegetation along the water, particularly leaf fall.

There are also human sources of organic material. When these are identifiable points of discharge into rivers and lakes, they are called *point* sources. Point sources of organic pollution include: (1) pulp and paper mills; (2) meat-packing plants; (3) food processing industries; and (4) wastewater treatment plants.

Nonpoint pollution comes from many sources that are difficult to identify. Nonpoint sources of organic pollution include:

1. Urban runoff of rain and melting snow that carries sewage from illegal sanitary sewer connections into storm drains; pet wastes from streets and sidewalks; nutrients from lawn fertilizers; and leaves, grass clippings, and paper from residential areas

2. Agricultural runoff that carries nutrients, like nitrogen and phosphates from fields

3. Runoff from animal feedlots that carries fecal material into rivers

BOD

Changes in Aquatic Life

In rivers with high BOD levels, much of the available dissolved oxygen is consumed by aerobic bacteria, robbing other aquatic organisms of the oxygen they need to live. Organisms that are more tolerant of lower dissolved oxygen may appear and become numerous, such as carp, midge larvae, and sewage worms. Organisms that are intolerant of low oxygen levels, such as caddisfly larvae, mayfly nymphs, and stonefly nymphs, will not survive. As organic pollution increases, the ecologically stable and complex relationships present in waters containing a high diversity of organisms is replaced by a low diversity of pollution-tolerant organisms.

Sampling Procedure

A dissolved oxygen bottle strapped to the extended rod sampler can be used to take a BOD sample. Remember, samples taken near the river bottom may hold more oxygen-demanding materials and organisms; therefore, to get a representative sample it is best to sample between the surface and river bottom, and away from the shore.

One of the dissolved oxygen bottles should be blackened, or purchased as a "dark bottle." One approach is to wrap the bottle with black electrical tape. It is always a good idea if several bottles are available to run several BOD samples.

Like the dissolved oxygen tests, it is important to run all tests for comparison at the same time of day.

Biochemical Oxygen Demand Testing Procedures

1. Fill two DO bottles (one clear and one black) with sample water, holding them for two to three minutes between the surface and the river bottom. If sampling by hand remember to use gloves.

2. Prepare the clear sample bottle according to the directions for the dissolved oxygen test. Determine the DO value for this sample in mg/L.

3. Place the black sample bottle in the dark and incubate for five days at 68°F (20°C). This is very close to room temperature in many buildings. If there is no incubator, place the blackened sample bottle in a "light-tight" drawer or cabinet.

4. After five days, determine the level of dissolved oxygen (in mg/L) of this sample by repeating steps four through eleven of the DO testing procedure.

5. The BOD level is determined by subtracting this DO level from the DO level found in the original sample taken five days previously:

   ➤ BOD = mg/L DO (original sample) - mg/L DO (after incubation)

   ➤ The BOD measure is, the amount of oxygen consumed by organic matter and associated microorganisms in the water over a five-day period.

BOD

Figure 3.14. Student viewing the precipitate in the biochemical oxygen demand test.

In waters suspected of carrying large amounts of organic waste/ sewage, the oxygen demand may be so great that all oxygen is consumed before the 5-day period. The above approach would not reveal the true oxygen demand over the 5-day period.

Alternative approaches require the use of a dissolved oxygen meter to period-ically measure dissolved oxygen levels, and re-saturate the sample with oxy-gen. Another alternative is to make buffered dilution water and dilute the sam-ple until oxygen demand is more in balance with oxygen supply.

For further information about testing procedures, please consult *Standard Methods For The Examination Of Water And Wastewater,* 16th edition, American Public Health Association, New York, 1985.

Temperature

The water temperature of a river is very important for water quality. Many of the physical, biological, and chemical characteristics of a river are directly affected by temperature. For example, temperature influences the:

1. Amount of oxygen that can be dissolved in water
2. Rate of photosynthesis by algae and larger aquatic plants
3. Metabolic rates of aquatic organisms
4. Sensitivity of organisms to toxic wastes, parasites, and diseases

Remember, cool water can hold more oxygen than warm water, because gases are more easily dissolved in cool water.

Human-Caused Changes in Temperature

TEMP

One of the most serious ways that humans change the temperature of rivers is through thermal pollution. Thermal pollution is an increase in water temperature caused by adding relatively warm water to a body of water. Industries, such as nuclear power plants, may cause thermal pollution by discharging water used to cool machinery. Thermal pollution may also come from stormwater running off warmed urban surfaces, such as streets, sidewalks, and parking lots.

People also affect water temperature by cutting down trees that help shade the river, exposing the water to direct sunlight.

Soil erosion can also contribute to warmer water temperatures. As discussed in Chapter 7, soil erosion can be caused by many types of activities, including the removal of streamside vegetation, overgrazing, poor farming practices, and construction. Soil erosion raises water temperatures because it increases the amount of suspended solids carried by the river, making the water cloudy (turbid). Cloudy water absorbs the sun's rays, causing water temperature to rise.

Changes in Aquatic Life

As water temperature rises, the rate of photosynthesis and plant growth also increases. More plants grow and die. As plants die, they are decomposed by bacteria that consume oxygen. Therefore, when the rate of photosynthesis is increased, the need for oxygen in the water (BOD) is also increased.

The metabolic rate of organisms also rises with increasing water temperatures, resulting in even greater oxygen demand. The life cycles of aquatic insects tend to speed up in warm water. Animals that feed on these insects can be negatively affected, particularly birds that depend on insects emerging at key periods during their migratory flights.

Most aquatic organisms have adapted to survive within a range of water temperatures. Some organisms prefer cooler water, such as trout and stonefly nymphs, while others thrive under warmer conditions, such as carp and dragonfly nymphs. As the temperature of a river increases, cool water species will be replaced by warm water organisms. Few organisms can tolerate extremes of heat or cold.

Temperature also affects aquatic life's sensitivity to toxic wastes, parasites, and disease. For example, thermal pollution may cause fish to become more vulnerable to disease, either due to the stress of rising water temperatures or the resulting decrease in dissolved oxygen.

Sampling Procedure

The temperature test measures the change in water temperature between two points: the test site and a site one mile upstream. By detecting significant temperature changes along a section of the river, we can begin to uncover the sources of thermal pollution.

TEMP

Because the temperature test compares the difference in water temperature at two different stream sites, it is important to match as closely as possible the physical conditions at these sites—current speed, amount of sunlight reaching the water, and the depth of the stream.

To reduce errors, the same thermometer should be used at both sites. Rubber gloves should be worn if there is any chance that hands might come in contact with the water.

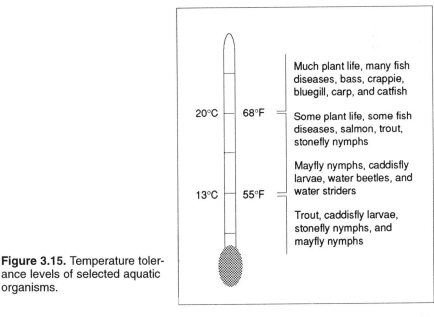

20°C — 68°F

Much plant life, many fish diseases, bass, crappie, bluegill, carp, and catfish

Some plant life, some fish diseases, salmon, trout, stonefly nymphs

Mayfly nymphs, caddisfly larvae, water beetles, and water striders

13°C — 55°F

Trout, caddisfly larvae, stonefly nymphs, and mayfly nymphs

Figure 3.15. Temperature tolerance levels of selected aquatic organisms.

Temperature Testing Procedure

1. At the site where the other water quality tests are being performed, lower the thermometer four inches below the water surface.

2. Keep the thermometer in the water until a constant reading is attained (approximately two minutes).

3. Record your measurement in Celsius. (See Figure 3.16 to convert Fahrenheit to Celsius.)

4. Repeat the test approximately one mile upstream as soon as possible.

$$°C = \frac{(°F - 32.0)}{1.80}$$

$$°F = (°C \times 1.80) + 32.0)$$

Figure 3.16. Temperature conversion chart.

5. Subtract the upstream temperature from the temperature downstream using the following equation:

temp. downstream - temp. upstream = temp. change

6. Record the change in temperature.

Total Phosphate (PO⁻₄-P)

TP

Phosphorus is usually present in natural waters as phosphate (PO^-_4-P). Organic phosphate is a part of living plants and animals, their by-products, and their remains. Inorganic phosphates include the ions ($H_2PO^-_2$, $HPO^=_4$, and PO^-_4) bonded to soil particles, and phosphates present in laundry detergents.

Phosphorus is an essential element for life. It is a plant nutrient needed for growth, and a fundamental element in the metabolic reactions of plants and animals. Plant growth is limited by the amount of phosphorus available. In most waters, phosphorus functions as a "growth-limiting" factor because it is usually present in very low concentrations.

The natural scarcity of phosphorus can be explained by its attraction to organic matter and soil particles. Any unattached or "free" phosphorus, in the form of inorganic phosphates, is rapidly taken up by algae and larger aquatic plants. Because algae only require small amounts of phosphorus to live, excess phosphorus causes extensive algal growth called "blooms." Algal blooms are a classic symptom of *cultural eutrophication*.

Cultural eutrophication is the human-caused enrichment of water with nutrients, usually phosphorus. Most of the eutrophication occurring today is human-caused. Natural eutrophication also takes place, but it is insignificant by comparison. Phosphorus from natural sources generally becomes trapped in bottom sediments or is rapidly taken up by aquatic plants. Forest fires and fallout from volcanic eruptions are natural events that cause eutrophication. Lakes that receive no inputs of phosphorus from human activities age very slowly.

Sources of Phosphorus

Phosphorus comes from several sources: human wastes, animal wastes, industrial wastes, and human disturbance of the land and its vegetation.

Sewage from wastewater treatment plants and septic tanks is one source of phosphorus in rivers. Sewage effluent should not contain more than 1 mg/L phosphorus according to the U.S. Environmental Protection Agency, but outdated wastewater treatment plants often fail to meet this standard. Also, some types of industrial wastes interfere with the removal of phosphorus at wastewater treatment plants.

Storm sewers sometimes contain illegal connections to sanitary sewers. Sewage from these connections can be carried into waterways by rainfall and melting snow. Phosphorus-containing animal wastes sometimes find their way into rivers and lakes in the runoff from feedlots and barnyards.

Soil erosion contributes phosphorus to rivers. The removal of natural vegetation for farming or construction, for example, exposes soil to the eroding action of rain and melting snow. Soil particles washed into waterways contribute more phosphorus.

Fertilizers used for crops, lawns, and home gardens usually contain phosphorus. When used in excess, much of the phosphorus in these fertilizers eventually finds its way into lakes and rivers.

Draining swamps and marshes for farmland or shopping malls releases nutrients like phosphorus that have remained dormant in years of accumulated organic deposits. Also, drained wetlands no longer function as filters of silt and phosphorus, allowing more runoff—and phosphorus—to enter waterways.

Impacts of Cultural Eutrophication

Shallow lakes and impounded river reaches, where the water is shallow and very slow-moving, are most vulnerable to the effects of cultural eutrophication. Phosphorus stimulates the growth of rooted aquatic vegetation. These plants, in turn, draw phosphorus previously locked within bottom sediments and release it into the water, causing further eutrophication. Eventually, the entire lake or river stretch may fill with aquatic vegetation.

The first symptom of cultural eutrophication is an algal bloom that colors the water a pea-soup green. As eutrophication increases, algal blooms become more frequent. Aquatic plants that normally grow in shallow waters become very dense. Swimming and boating may become impossible.

The advanced stages of cultural eutrophication can produce anaerobic conditions in which oxygen in the water is completely depleted. These conditions usually occur near the bottom of a lake or impounded river stretch, and produce gases like hydrogen sulfide, unmistakable for its "rotten egg" smell.

Changes in Aquatic Life

As with other types of water pollution, cultural eutrophication causes a shift in aquatic life to a fewer number of pollution tolerant species. The many different species that exist in clean water are replaced by a fewer number of species that can tolerate low dissolved oxygen levels—carp, midge larvae, sewage worms (Tubifex), and others. For example, waters that once supported bass, walleye, pike, and bluegill may only be able to support carp under eutrophic conditions.

Reversing the Effects of Cultural Eutrophication

Aquatic ecosystems have the capacity to recover if the opportunity is provided through the following practices:

TP

1. Reducing our use of lawn fertilizers (particularly inorganic forms) that drain into waterways

2. Encouraging better farming practices: low-till farming to reduce soil erosion; soil-testing to match the amount of fertilizer applied to soil needs, thus preventing excess fertilizer from finding its way into waterways; and building storage or collecting areas around cattle feedlots, so that phosphorus containing manure is not carried away with surface runoff

3. Preserving natural vegetation whenever possible, particularly near shorelines; preserving wetlands to absorb nutrients and maintain water levels; and enacting strict ordinances to prevent soil erosion

4. Supporting measures (including taxes) to improve phosphorus removal by wastewater treatment plants and septic systems; treating storm sewer wastes if necessary; and encouraging homeowners along lakes and streams to invest in community sewer systems

5. Requiring particular industries to pretreat their wastes before sending it to a wastewater treatment plant

Can you think of any other actions that would prevent or reduce the effects of eutrophication?

Sampling Procedure

It is important that glassware used for measuring total phosphate be "acid-washed"; that is, soaked in dilute HCL and then rinsed thoroughly with distilled water. Please wear protective gloves when handling this glassware. *WARNING: Never wash this glassware with phosphorus-containing detergents.*

Total Phosphate (PO$^-_4$-P) Testing Procedure

1. Fill the 50 mL graduated cylinder to the 50 mL line with the water sample. Pour into a 125 mL Erlenmeyer flask. Use gloves if drawing the sample by hand.

2. Use a 1 mL pipet to add 1 mL of Sulfuric Acid, 36% to the flask. Swirl to mix.

3. Use the 0.05 g spoon to add one measure of Ammonium Persulfate. Swirl to dissolve.

4. Add a few boiling stones. Place the flask on a hot plate, small backpacking stove or Sterno and boil gently for 30 minutes. Add deionized water to the sample during the boiling to maintain a volume between 10 and 50 mL. Permit the volume to decrease to approximately 10 mL (about 1/4 inch of water) at the end of the boiling step but do not allow the sample to go to dryness or to dense white sulfur trioxide fumes. Remove from the hot plate and cool.

   If inside, please boil sample in a well-ventilated place; if outside, please stay upwind of the boiling sample.

5. Add one drop of Phenolphthalein Indicator, 1% to the cooled sample.

6. While swirling the flask, use a 1 mL pipet to add Sodium Hydroxide dropwise until the solution turns faint pink. A volume of slightly less than 3 mL is required.

7. While swirling the flask, add Sulfuric Acid, 36%, one drop at a time, until the pink color disappears.

8. Quantitatively transfer the sample, which should be at room temperature, to the 50 mL graduated cylinder. After transferring the solution from the flask to the graduated cylinder, wash the flask with a little deionized water and add it to the solution in the graduated cylinder. Dilute the solution in the graduated cylinder to exactly 50 mL using deionized water and mix well.

TP

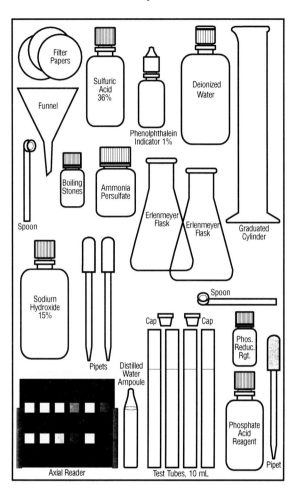

Figure 3.17. Total Phosphate test kit items.

Orthophosphate Test (low range, 0–2 mg/L phosphate)

9. Fill a test tube to the 10 mL line with the test sample from step 9.

10. Use the 1.0 mL pipet to add 1.0 mL of Phosphate Acid Reagent. Cap and mix.

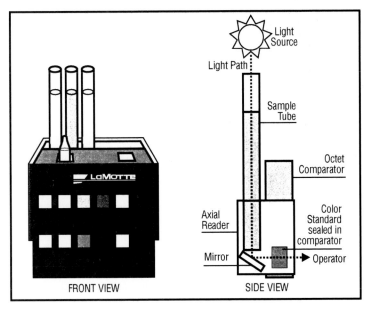

Figure 3.18. Use of the Axial Reader

11. Use the 0.1 g spoon to add one level measure of Phosphate Reducing Reagent. Cap and mix until the powder has dissolved. Wait 5 minutes.

12. Remove the stopper from the test tube. Place the tube in the Phosphate Comparator with Axial Reader. Match the sample color to a color standard. Record the result as mg/L (ppm) Total Phosphate.

> **Note:**
>
> *Total phosphate concentrations of non-polluted waters are usually less than 0.1 mg/L.*

TP

Nitrates (NO$^-_3$)

Nitrogen is an element needed by all living plants and animals to build protein. In aquatic ecosystems, nitrogen is present in many forms.

Nitrogen is a much more abundant nutrient than phosphorus in nature. It is most commonly found in its molecular form (N_2), which makes up 79 percent of the air we breathe. This form, however, is useless for most aquatic plant growth.

Blue-green algae, the primary algae of algal blooms, are able to use N_2 and convert it into forms of nitrogen that plants can take up through their roots and use for growth: ammonia (NH_3) and nitrate (NO^-_3).

How do aquatic animals obtain the nitrogen they need to form proteins? In two ways: they either eat aquatic plants and convert plant proteins to specific animal proteins, or they eat other aquatic organisms which feed upon plants.

As aquatic plants and animals die, bacteria break down large protein molecules into ammonia. Ammonia is then oxidized (combined with oxygen) by specialized bacteria to form nitrites (NO^-_2) and nitrates (NO^-_3). These bacteria get energy for metabolism from oxidation.

Excretions of aquatic organisms are very rich in ammonia, although the amount of nitrogen they add to waters is usually small. Duck and geese, however, contribute a heavy load of nitrogen (from excrement) in areas where they are plentiful. Through decomposition of dead plants and animals, and the excretions of living animals, nitrogen that was previously "locked up" is released.

There even exist bacteria that can transform nitrates (NO^-_3) into free molecular nitrogen (N_2). The nitrogen cycle begins again if this molecular nitrogen is converted by blue-green algae into ammonia and nitrates.

Because nitrogen, in the form of ammonia and nitrates, acts as a plant nutrient, it also causes eutrophication. As you learned in the Total Phosphate section, eutrophication promotes more plant growth and decay, which in turn increases biochemical oxygen demand. However, unlike phosphorus, nitrogen rarely limits plant growth, so plants are not as sensitive to increases in ammonia and nitrate levels.

Sources of Nitrates

Sewage is the main source of nitrates added by humans to rivers. Sewage enters waterways in inadequately treated wastewater from sewage treatment plants, in the effluent from illegal sanitary sewer connections, and from poorly functioning septic systems.

Septic systems are common in rural areas. Unlike large, centralized urban sewer systems that collect waste from many households, septic systems are generally used to treat the waste from only a single household.

In a septic system, household wastewater from toilets, sinks, bathtubs, and washing machines flows through a main pipe into a box called a septic tank. After larger waste materials settle and floating grease is skimmed off, the remaining liquid then flows through a grid of perforated pipes. The holes in these pipes allow the liquid to trickle out onto a layer of stone, gravel, and soil known as the "drain field". (See Septic Systems, page 154, for more information about septic systems.)

In properly functioning septic systems, soil particles remove nutrients like nitrates and phosphates before they reach groundwater. However, two factors often keep septic systems from working like they should.

Septic systems must be properly located. When septic system drainfields are placed too close to the water table, nutrients and bacteria are able to percolate down into the groundwater where they may contaminate drinking water supplies. They may also find their way into lakes or rivers via groundwater flow.

Also, septic tanks must be emptied periodically to function properly. If the tank is full, household wastes go directly to the drain field instead of settling in the tank. When this happens, the drain field pipes may become plugged, and household sewage may start to pool on the ground and enter water through surface runoff.

Water containing high nitrate levels can cause a serious condition called methemoglobinemia (met-hemo-glo-bin-emia), if it is used for infant milk formula. This condition prevents the baby's blood from carrying oxygen; hence the nickname "blue baby" syndrome.

Two other important sources of nitrates in water are fertilizers, and the runoff from cattle feedlots, dairies, and barnyards. High nitrate levels have been discovered in groundwater beneath croplands due to excessive fertilizer use, especially in heavily irrigated areas with sandy soils. Stormwater runoff can carry nitrate-containing fertilizers from farms and lawns into waterways. Similarly, places where animals are concentrated, such as feedlots and dairies, produce large amounts of wastes rich in ammonia and nitrates. If not properly contained, these can seep into groundwater or be transported in runoff into surface waters.

As discussed in the Total Phosphate section, people have created the eutrophication problem that threatens to limit organism diversity, recreational opportunities, and property values. Only we can reverse eutrophication through thoughtful action.

Figure 3.19. Some nitrate kits will provide accurate readings in both brackish and fresh waters, while others only test accurately in fresh water.

Sampling Procedure

NO$_3$

Again, any sampling device might be used for this water quality test to obtain representative samples. It is also important to have spotless glassware, rinsed with demineralized water. *Always use demineralized water during the nitrate test.* Distilled water contains ammonia (NH$_3$) ions that will interfere with the test.

Nitrate Testing Procedure

Note:

Use the following procedure for suspected nitrate nitrogen in the 0.25–10.0 mg/L range. A low range test 0–1 mg/L Nitrate test is also available; see the test equipment section in Appendix A.

---

1. Fill the sample bottle with sample water. Use gloves if drawing the sample by hand.
2. Rinse and fill one test tube to the 2.5 mL line with water from the sample bottle.

3. Dilute to the 5 mL line with the Mixed Acid Reagent. Cap and mix. Wait 2 minutes.

4. Use the 0.1 g spoon to add one level measure (avoid any excess) of Nitrate Reducing Reagent. Cap and invert gently 50–60 times in one minute. Wait 10 minutes.

5. Insert the test tube into the Nitrate Nitrogen Comparator. Match the sample color to a color standard. Record the result as mg/L (ppm) Nitrate Nitrogen (NO_3-N). To convert to mg/L Nitrate (NO_3), multiply by 4.4.

6. Place the reacted sample in a clearly marked container. Arrangements should be made with toxic material handlers for safe disposal. Please wash your hands after this water test is completed.

For Your Information

Nitrate Test Kits which do not require the use of cadmium are also available. For ordering information see Appendix A.

NO_3

Figure 3.20. Cadmium waste should be rinsed into a clearly marked container.

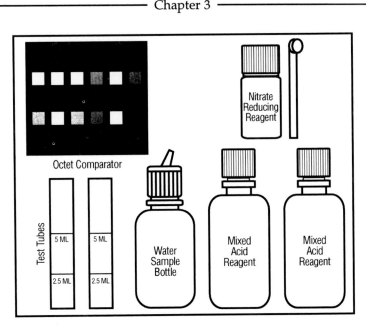

Figure 3.21. Nitrate test kit items.

Turbidity

Turbidity is a measure of the relative clarity of water: the greater the turbidity, the murkier the water. Turbidity increases as a result of suspended solids in the water that reduce the transmission of light. Suspended solids vary: ranging from clay, silt, and plankton, to industrial wastes and sewage.

TURB

Sources of Turbidity

High turbidity may be caused by soil erosion, waste discharge, urban runoff, abundant bottom feeders (such as carp) that stir up bottom sediments, or algal growth. The presence of suspended solids may cause color changes in water, from nearly white to red-brown, or to green from algal blooms.

Changes in Aquatic Life

At higher levels of turbidity, water loses its ability to support a diversity of aquatic organisms. Waters become warmer as suspended particles absorb heat from sunlight, causing oxygen levels to fall. (Remember, warm water holds less oxygen than cooler water.) Photosynthesis decreases because less light penetrates the water, causing further drops in oxygen levels. The combination of warmer water, less light, and oxygen depletion makes it impossible for some forms of aquatic life to survive.

Suspended solids affects aquatic life in other ways. Suspended solids can clog fish gills, reduce growth rates, decrease resistance to disease, and prevent egg and larval development. Particles of silt, clay, and organic materials settle to the bottom, especially in slow-moving stretches of rivers. These settled particles can smother the eggs of fish and aquatic insects, as well as suffocate newly-hatched insect larvae. Material that settles into the spaces between rocks makes these microhabitats unsuitable for mayfly nymphs, stonefly nymphs, caddisfly larvae, and other aquatic insects living there.

Sampling Procedures

Turbidity can be measured using a simple device called a Secchi disk, or a more precise instrument known as a turbidimeter.

A Secchi disk is an 8" diameter (23 cm.) black and white disk attached by a chain or rope that is marked in foot or meter increments. Because Secchi disk measurements are based upon the disk being lowered until it disappears, it cannot be used in rivers which are shallow or have low turbidity. In these cases the Secchi disk reading may need to be estimated as accurately as possible.

It may be difficult to use the Secchi disk in fast river currents because the current will push the disk downriver, preventing an accurate measurement. A weight may have to be added to the disk in this situation.

A turbidimeter (Figure 3.22) is an optical device that measures the scattering of light, and provides a relative measure of turbidity in nephelometer turbidity units (NTUs). Secchi disk measurements and turbidimeter results can be roughly equated.

TURB

Turbidimeters are relatively expensive, but are the most accurate method for measuring turbidity. If multiple schools are involved in a watershed monitoring program, then one turbidimeter might be purchased for the program. Samples from throughout the watershed could then be read on the one turbidimeter. It may also be possible to use a local college's turbidimeter.

Another option is to use an inexpensive turbidity test available from the LaMotte Company. Like the original Jackson Tube, this test involves viewing an object located at the end of a tube; in this case, the object is a black dot rather than a flame. As the turbidity of a sample increases, the dot becomes increasingly blurred. The turbidity of the sample is then compared with an identical amount of clear water to which a standardized turbidity reagent has been added. See Appendix A for ordering information.

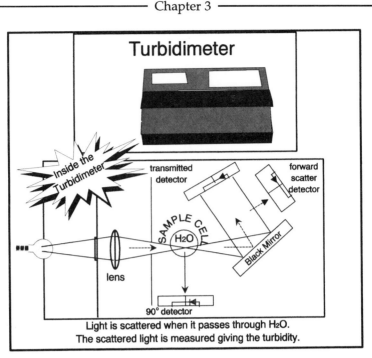

Figure 3.22. An internal drawing of a turbidimeter showing the passage of light and its scatter.

Turbidity Testing Procedures: Secchi Disk

TURB

1. Lower the Secchi disk from a bridge, boat, or dock into the water until it disappears. It is important that the disk travels vertically through the water and is not "swung out" by the river current. Note the number of feet/inches on the chain or rope.

2. Drop the disk even further (until it disappears) and then raise it until you can see the disk again. Note the number of feet/inches on the chain.

3. Add the results of step 1 and step 2 and divide by two. This is your turbidity level using the Secchi disk.

The Relationship Among Feet, JTU's, and NTU's

The Secchi disk measurement in feet has been roughly correlated with Jackson Turbidity Units (JTU's). These units were based upon a standard suspension of 1000 parts per million diatomaceous earth in water. By diluting this suspension a series of standards was produced.

Jackson Turbidity Units (JTU's) are the application of these standards to the original device for measuring turbidity called the "Jackson tube". The Jackson tube is a long glass tube suspended over a lit candle. A sample of water was slowly poured into the tube until the candle flame as viewed from above became diffuse. This device is no longer used because it is not sensitive to very low turbidities.

A turbidimeter measures turbidity as nephelometer turbidity units (NTU). Instruments such as the turbidimeter that measure the scattering of light are called nephelometers. Both NTU's and JTU's are interchangeable units. They differ only in that their name reflects the device used to measure turbidity.

Please note that the weighted curve chart in Chapter 4 uses NTU's/JTU's. Because feet have been roughly correlated with JTU's, the numerical value obtained from the weighted curve chart should be viewed cautiously when using a Secchi disk. The weighting factor for turbidity, however, is quite small (0.08) and does not affect the overall water quality index as heavily as most of the other water quality parameters.

Note:

This is a comparative figure and the results can be transferred to the curve chart in Chapter 4. It does provide a value that can be compared with previous readings.

TURB

Total Solids

This water quality measure (also referred to as total residue) includes: (1) dissolved solids, or that portion of the solid matter found in a water sample that passes through a filter; and, (2) suspended solids, or that portion of solid matter that is trapped by a filter.

As discussed in the turbidity section, suspended solids include anything from silt and plankton, to industrial wastes and sewage.

Dissolved or inorganic materials include calcium, bicarbonate, nitrogen, phosphorus, iron, sulfur, and other ions found in a water body. A constant level of these materials is essential for the maintenance of aquatic life because the density of total solids determines flow of water in and out of an organism's cells. Also, many of these dissolved ions, such as nitrogen, phosphorus, and sulfur, are building blocks of molecules necessary for life.

Sources of Total Solids

Many sources can affect the natural balance of total solids in waterways. One example is runoff from urban areas, which can carry salt from streets in winter, fertilizers from lawns, and many other types of material. Another source are wastewater treatment plants, which can add phosphorus, nitrogen, and organic matter to rivers.

Some examples of sources of suspended solids include leaves and other plant materials, soil particles from urban runoff and soil erosion, and decayed plant and animal matter that is converted into particulate matter within the river.

Changes in Aquatic Life

High concentrations of total solids can lower water quality and cause water balance problems for individual organisms. On the other hand, low concentrations may limit the growth of aquatic life. Phytoplankton, for example, are totally dependent upon nitrates and phosphates that are dissolved in the water.

High concentrations of dissolved solids can lead to laxative effects in drinking water and impart an unpleasant mineral taste to the water. As discussed in the turbidity section, high concentrations of suspended solids also: (1) reduce water clarity; (2) contribute to a decrease in photosynthesis; (3) bind with toxic compounds and heavy metals; and (4) lead to an increase in water temperature through greater absorption of sunlight by surface waters.

TS

Sampling Procedures

As mentioned before, it is important to obtain representative water samples either from the middle of the river or at least as far out from the shore as possible. Water samples should also be taken between the surface and river bottom if possible. The extended rod sampler could be used to obtain water samples from shore.

If a sampling device is not available, reach from shore as far as possible, being careful not to disturb the river bottom. If hands will come in contact with the water, make sure to wear gloves.

Total Solids Testing Procedures

1. Place a glass-stoppered bottle (one that holds at least 100 mL) about halfway to the bottom of the river. (Remember those gloves!) Open the bottle and fill. Put the stopper on the bottle and remove from the water. Remove any large floating particles or submerged masses from the sample.

2. In the laboratory, clean a 300 mL beaker (a 300 mL beaker provides greater surface area) and dry in a 103°C oven for one hour. The beaker may also be placed over a burner with low red heat.

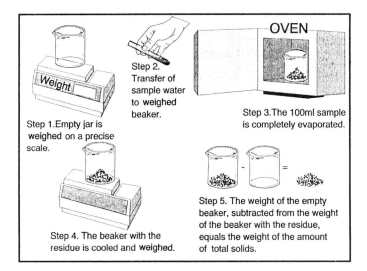

Step 1. Empty jar is weighed on a precise scale.

Step 2. Transfer of sample water to weighed beaker.

Step 3. The 100ml sample is completely evaporated.

Step 4. The beaker with the residue is cooled and weighed.

Step 5. The weight of the empty beaker, subtracted from the weight of the beaker with the residue, equals the weight of the amount of total solids.

TS

Figure 3.23. Steps for measuring total solids.

3. Remove beaker from heat with tongs and allow it to cool, then weigh with a sensitive balance (to the nearest .0001 gram). *Do not touch the beaker with bare hands because body moisture will be transferred to it, thereby changing its weight.* Use tongs, if available, or pads or gloves.

4. Using a pipette or graduated cylinder, measure a 100 mL sample and transfer into the 300 mL beaker. If sample has been sitting, swirl the sample water before measuring out the 100 mL.

5. Place the 300 mL beaker with the sample in a 103°C oven overnight to evaporate the liquid and dry the resulting residue. Allow the beaker to cool, then reweigh it. *Remember: don't touch the beaker with your hands.*

6. Subtract the initial weight (in grams) of the empty beaker from the weight of the beaker and residue to obtain the increase in weight, or the weight of the residue. (See formula below.)

The formula for determining total solids is:

$$\frac{\text{Increase in weight in gm}}{\text{Volume in millileters (mL)}} \times \frac{1000 \text{ mg.}}{1 \text{ gram}} \times \frac{1000 \text{ mL.}}{1 \text{ liter}} = \text{mg/L}$$

EXAMPLE: Weight of beaker and residue = 48.2982 grams

Weight of beaker   48.2540 grams
Weight of residue   .0442 grams

$$\frac{.0442 \text{ grams}}{100 \text{ mL}} \times \frac{1.000 \text{ mg.}}{1 \text{ gram}} \times \frac{1.000 \text{ mL.}}{1 \text{ liter}} = 442 \text{ mg/L}$$

TS

Figure 3.24. Students reviewing the water quality data and using hand computers to calculate the overall water quality index (WQI).

GREEN Low Cost Water Monitoring Kit

This low cost kit from the LaMotte Company is packaged in an unbreakable plastic canister and can be purchased for US $27. It includes all the needed equipment, reagents, instructions, and safety guidelines to test water for: dissolved oxygen, total coliform bacteria, pH, biochemical oxygen demand, temperature change, nitrates, phosphates, and turbidity. The chemicals are non-hazardous tabletized reagents. For ordering information, see page 221. Price: $27.00

TS

GREEN Advanced Water Monitoring Kit

From the LaMotte Company, this kit contains the basic equipment for water monitoring as described in this chapter. This large kit includes the following kits: dissolved oxygen, total coliform bacteria, pH, temperature change, biochemical oxygen demand, nitrates, phosphates, and turbidity. It also includes an armored thermometer (non-mercury), and bottles for testing biochemical oxygen demand. The kit also includes benthic macroinvertebrate study material. For ordering information, see page 221. Price: $159.00

CHAPTER

Calculating the Results

Procedure for Calculating the Overall Water Quality Index

After the nine water quality tests are completed and the results of each test recorded, the Water Quality Index (WQI) for the section (or sections) of the river monitored can be computed. To formulate a WQI, you must first compute Q-values for the results you obtained for each of the nine tests. To do that, follow these steps for each test:

1. Find the weighting curve chart for that test in this chapter.

2. Locate your test result on the bottom (horizontal or "x" axis) of the chart.

3. Interpolate the Q-value for your test result using the following steps:

   a. From your test result value on the horizontal ("x") axis of the chart, draw a vertical line up until it intersects the weighting curve line.

   b. From this point of intersection, draw a horizontal line to the left hand side (the vertical or "y" axis) of the chart.

   c. Where this horizontal line intersects the vertical ("y") axis of the chart, read off the value. This is the Q-value for this test; it should be recorded in Column B on the WQI chart (Chart 10) at the end of this chapter.

4. Repeat these steps to find the Q-values for each of the nine tests. Record the Q-values on Chart 10.

The Q-value for each test should then be multiplied by the weighting factor shown on Chart 10 for each test. Record the product of this calculation in Column D of the chart.

The weighting factor provides a measure of the relative importance of each test to overall water quality. For example, dissolved oxygen has a weighting factor of 0. 17, which means it is considered more important in determining overall water quality than total solids, which has a weighting factor of only 0.07.

To determine the overall Water Quality Index (WQI) for the section of the river you monitored, add the totals of the nine tests in Column D. How do your results for this stretch compare with past years? How about comparisons with other rivers in your region, nation, or globally? What did you learn about your river? Are there any actions that need to be taken to improve water quality, and thereby the lives of people that depend upon your river?

Example

Suppose you measured a fecal coliform count of 10,000 colonies per 100 ml. Turn to the weighted curve chart for fecal coliform (Chart 2), and locate 10,000 on the horizontal axis. Draw a vertical line up from this point until it intersects the curved line. From this point, draw a horizontal line over to the vertical axis of the chart. What do you get?

You should get 10. This is your Q-value for the fecal coliform test. You should enter this value in Column B of Chart 10, and then multiply it by the weighting factor shown in Column C for fecal coliform, which is 0.16. The result (1.6, in this case) is your total for this test, and should be recorded in Column D.

Weighting Curve Charts

Chart 1: Dissolved Oxygen (DO) Test Results

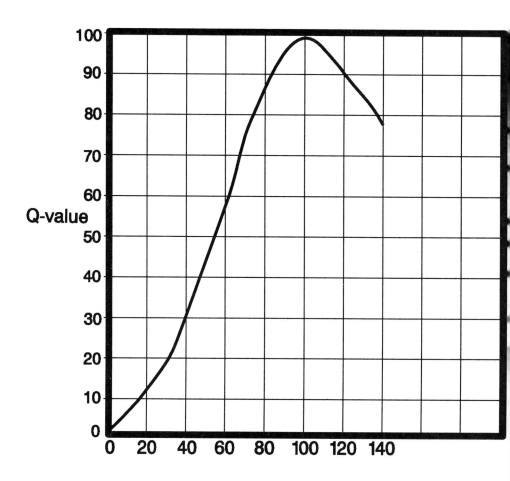

DO: % saturation

Note: if DO % saturation > 140.0, Q = 50.0

Chart 2: Fecal Coliform (FC) Test Results

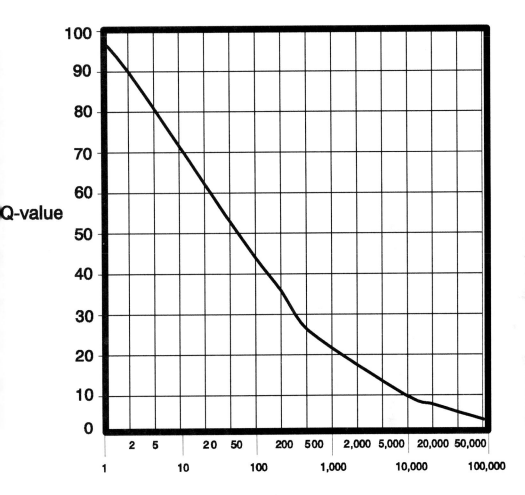

FC: colonies/100 mL

Note: if FC > 10^5, Q = 2.0

Chart 3: pH Test Results

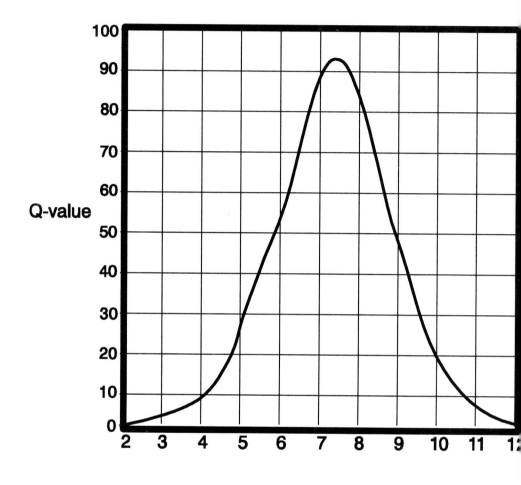

pH: units

Note: if pH , 2.0, Q = 0.0; if pH > 12.0, Q = 0.0

Chart 4: 5-Day Biochemical Oxygen Demand (BOD) Test Results

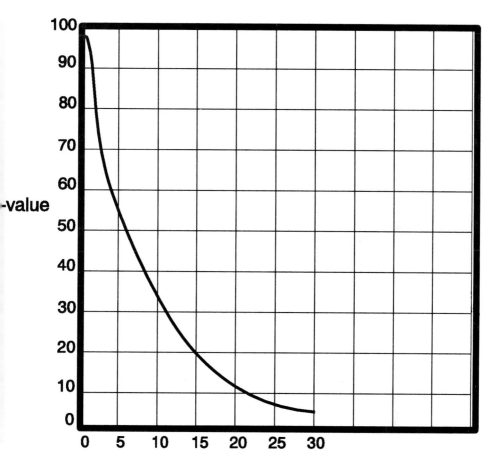

BOD$_5$: mg/L

Note: if BOD$_5$ > 30.0, Q = 2.0

Chart 5: Change in Temperature (ΔT, °C) Test Results

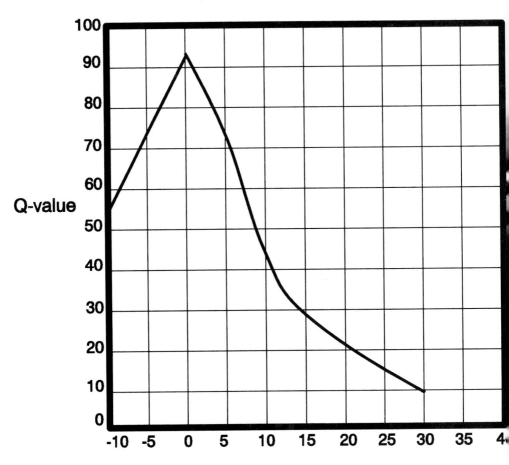

ΔT: °C

Chart 6: Total Phosphate (as PO$^-_4$-P) Test Results

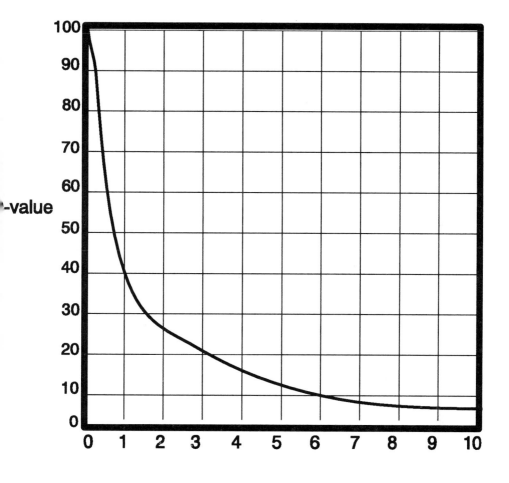

PO$^-_4$-P: mg/L

Note: if PO$^-_4$-P > 10.0, Q = 2.0

Chart 7: Nitrate (as NO⁻₃) Test Results

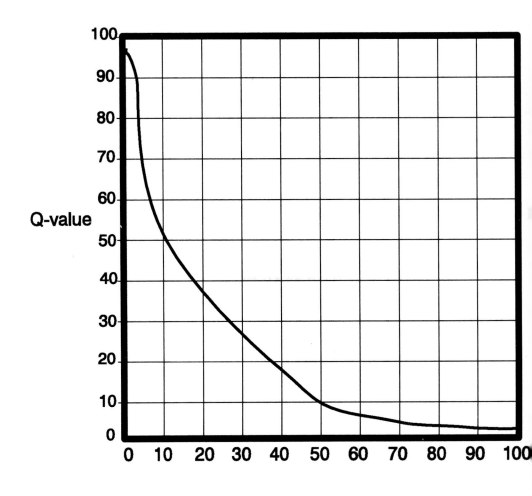

Q-value

NO⁻₃; mg/L

Note: if $NO^-_3 > 100.0$, Q = 1.0

Chart 8: Turbidity Test Results

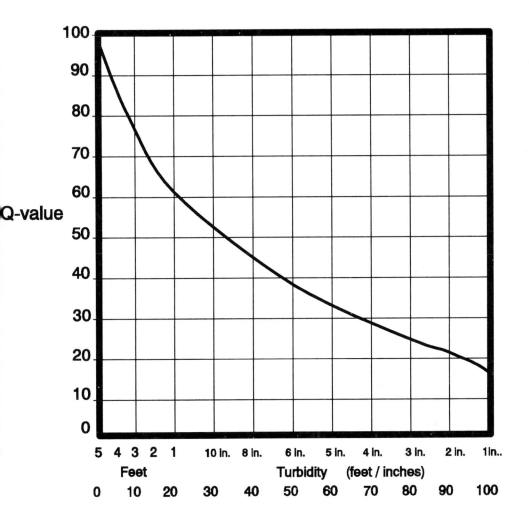

Q-value

Feet

Turbidity   (feet / inches)

Turbidity: NTU's/JTU's

Note: if Turbidity > 100.0, Q = 5.0

Chart 9: Total Solids (TS) Test Results

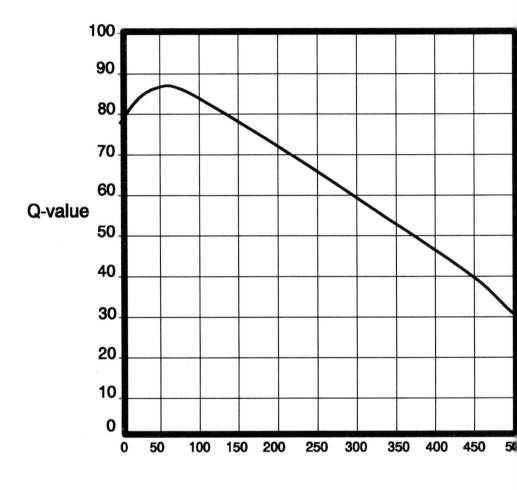

TS: mg/L

Note: if TS > 500.0, Q = 20.0

Chart 10: Calculating the Overall Water Quality
of a Section of a River System

Date _____ Time _____

Test Location _____

Weather Conditions_____

Water Tests	Text Page	Chart Page
Dissolved Oxygen	27	76
Fecal Coliform	34	77
pH	43	78
BOD	47	79
Temperature	51	80
Total Phosphate	54	81
Nitrates	60	82
Turbidity	66	83
Total Solids	70	84
Water Quality Index	74	85

	Test Results (Column A)	Q-Value (Column B)	Weighting Factor (Column C)	TOTAL (Column D)
1. DO	% sat.		0.17	
2. Fecal Coliform	colonies/100 mL		0.16	
3. pH	units		0.11	
4. BOD	mg/L		0.11	
5. Temperature	Δ°C		0.10	
6. Total Phosphate	mg/L		0.10	
7. Nitrates	mg/L		0.10	
8. Turbidity	NTU or Ft.		0.08	
9. Total Solids	mg/L		0.07	

Overall Water Quality Index _____

Assessment of Toxics Using a Bioassay

The nine water quality indicators described as a part of the Water Quality Index can only reveal organic forms of pollution or physical disturbances to a stream or river. Biological monitoring has the capacity to point to the possibility of toxic contamination of a river. Neither of these approaches, however, are designed to directly show the impact that a point source or sediment might have upon aquatic life. Analytical approaches that generate ppm or ppb values for specific toxics are too expensive and employ toxic chemicals. The assessment of toxics in aquatic environments embrace a constellation of approaches that include the use of living organisms to measure the toxic effects of water and sediments; these are known as bioassays. A lettuce seed bioassay is described in this chapter. But first let's look at two major forms of toxicants—inorganic and organic—their sources, ecological effects, and human health impacts.

Inorganic Toxicants

The term "heavy metals" is used interchangeably with "trace metals" and refers to the metallic elements of the periodic table. Metals are highly electropositive (yielding electrons easily). Some common metals include: Aluminum (A1), Arsenic (As), Cadmium (Cd), Chromium (Cr), Copper (Cu), Iron (Fe), Lead (Pb), Manganese (Mn), Mercury (Hg), Nickel (Ni), Silver (Ag), and Zinc (Zn).

In many waters of North America and around the world, bottom sediments are laden with trace metals. The Detroit River—bordering the United States and Canada—holds elevated levels of cadmium, chromium, copper, and zinc in its sediments. More than 30 percent of the bullheads in the lower Trenton Channel (Detroit River) have skin or liver tumors (International Joint Commission, 1987). These levels are attributed to the Rouge River, and to industrial and combined sewers.

In Minamata, Japan, years of dumping by the Chisso Corporation have left very high levels of mercury in Minamata Bay and methylmercury poisoning among many people. The lower Rhine River, Germany is plagued by high levels of cadmium, manganese, iron, zinc, and copper concentrations. These areas reflect metal contamination beyond background levels. It is easy to forget that these metals also have sources tied to natural processes and cycles.

Sources of Metals

Volcanic eruptions, weathering of rock, and other natural processes continually introduce and cycle metals in the environment. This geologic weathering is responsible for the background levels of metals found in rivers and lakes. Natural processes and cycles are disrupted by mining and manufacturing processes that redistribute and concentrate metals in the environment. Lead mines, silver mines, copper mines, and iron ore mines represent human activities that displace and transform these metals for other uses.

Common uses of heavy metals in manufacturing include: lead and nickel in batteries, copper in textiles, silver in photographic film, and iron ore in steel production. Sewer and wastewater treatment plant effluent, atmospheric deposition, and landfill leachate also carry concentrated metal levels to surface water and groundwater. These pathways represent both point and nonpoint sources of metal pollution.

Point and Nonpoint Sources

Point sources, like sewage effluent, may contain elevated levels of copper, lead, zinc, and cadmium. Some of this increase can be attributed to corrosion within the wastewater supply pipes (Forstner, 1979).

Nonpoint sources of pollution include both urban and rural runoff. Urban stormwater runoff carries with it increased metal loadings. Whipple and Hunter (1977) found high concentrations of lead, zinc, and copper within the first 30 minutes of a storm event; this time period is called the first flush. Stormwater carries lead deposited on streets and parking lots from car exhaust, oil, and grease; zinc in motor oil and grease; and, copper worn from metal plating and brake linings.

Rural runoff is primarily composed of sediment. Sediments eroding from croplands carry with them cadmium, and even uranium, which are both found in some phosphate fertilizers (Forstner, 1979). Herbicides used to control weeds may also contain arsenic.

Metals used in products common to our daily life, like cars, eventually end-up in landfills, or their by-products can be transported via stormwater to a river. Either path leads to significant ecological effects.

Ecological Effects of Metals

Trace metals include essential elements for growth and metabolism. Nickel, zinc, and copper are considered essential elements. Non-essential elements include cadmium, mercury, and lead. Essential trace metals needed for metabolism and growth may, at excessive levels, become toxic to invertebrates and fish. Often, the difference between non-toxic and toxic levels is minute.

For non-essential metals, like mercury and cadmium, even very low levels can be toxic to organisms. Toxicity refers to the potential harmful effects of a chemical upon a living organism. Harmful effects can be defined as death (lethal effects) or inability to reproduce. Some effects from exposure to toxic chemicals are more subtle—behavioral changes, or changes in growth and development (sub-lethal effects). It becomes difficult to unravel the many interconnected effects related to toxic metal levels. For example, a fish that is stressed by accumulation of metals may physically become less able to avoid predation.

The toxicity of heavy metals to aquatic organisms depends upon many factors, including the bioavailability of metals to organisms. Organisms uptake metals via the food they consume, through adsorption onto membranes (gills), and transport through the skin. Bioavailability in turn is influenced by water hardness, pH, life cycle and age of the organism, and water temperature.

With increasing water hardness, the toxicity of metals decreases, as they are adsorbed onto insoluble carbonate compounds. A lowering of pH increases the solubility of metals in solution. Below a pH of 5.5, aluminum levels may become a threat to aquatic life (Forstner, 1979). Mercury levels have also increased in lakes affected by acid rain. Concentrations of metals, like mercury, are often higher in older organisms. An increase in water temperature increases metabolism and quickens the intake of metals as well.

Metals are adsorbed onto organic material and so are found concentrated in bottom sediments. Researchers find that organisms that inhabit bottom sediments, like *Tubifex*, exhibit high levels of metals.

People who eat bottom-feeding fish like carp and bullhead on a frequent basis may be risking their health.

Public Health Effects of Metals

Some metals are essential to human health:

➤ nickel (needed for making blood cells)
➤ iron (forms an iron complex called hemoglobin in the blood)

➤ zinc (activator of enzymes)

➤ manganese (important in glucose utilization).

The absence of these metals can lead to deficiencies in growth and development. In excessive amounts, however, these same metals may become toxic. There are also metals, as mentioned before, that are non-essential (cadmium and lead) and toxic in even small amounts.

Certain populations of people are at greater risk of heavy metal contamination:

1. workers that handle toxic chemicals on a regular basis
2. communities that sit in a heavy industrial setting
3. people who eat fish regularly
4. communities with lead pipes or pipes with lead solder.

Historically, metals have been used in many different ways. In Roman times, lead lined the water ducts and cooking vessels. Lead poisoning may have contributed to the downfall of the Roman Empire. Lead has also been widely used in lead-based paints and in tin-lead solder used by plumbers until recently. Mercury was used in the manufacture of hats, from which the phrase, "mad as a hatter" was derived. This phrase referred to some of the physical symptoms of mercury poisoning among hat makers (loss of coordination, slurred speech, vision loss).

The Environmental Protection Agency (EPA) has recommended standards for human health, based upon body burden estimates. A body burden is the amount of toxin a person can withstand before becoming ill. Standards are derived from chronic toxicity tests on rats and other mammals. The EPA standards are shown in the chart on the next page.

Symptoms of metal poisoning vary with specific metals and with specific metal "species" or forms. For example, the most dangerous form of mercury is methylmercury. Methylmercury is formed by the methylation of inorganic mercury by bacteria. The organic form of mercury is more readily taken up by fat in the body and can cause Minamata disease. Symptoms of methylmercury contamination include: slurred speech, loss of feeling in extremities, loss of coordination, blindness, and even death.

Tragic consequences to human health from metal poisoning can be reduced through systematic monitoring. Heavy metal monitoring is important to evaluate whether water quality standards are being achieved, and to diagnose formerly hidden threats to ecosystem health.

Organic Toxicants

Unlike the inorganic metals, these toxicants can only be manufactured or produced indirectly. There are literally dozens of books and thousands of research papers devoted to understanding these toxicants more completely. Included here is a very brief description of the major groups.

Polycyclic Aromatic Hydrocarbons (PAHs)

Polycyclic aromatic hydrocarbons (PAHs) consist of just carbon and hydrogen atoms, in the form of 2 or more fused benzene rings. There are two principal sources of PAHs to the environment; the incomplete combustion of organic matter, and spills or other discharges of petroleum products, which contain high concentrations of PAHs (up to 7% by weight in some crude oils, and 15% in synthetic oils). Certain industrial processes can also be significant, local sources of PAHs; these include coal coaking, creosote and coal tar production, and waste incineration. Significant concentrations of PAHs are also found in urban storm water runoff, from tire particles, condensed vehicle exhaust, and leaching from asphalt road surfaces. Tobacco smoke is a major source of PAHs for some people. There is usually a direct correlation between the concentration of total PAHs in water and the degree of industrial development in a watershed. Unpolluted fresh or salt water usually contains less than 0.1 ppb of total PAHs, while urban river waters typically contains 1 to 5 ppb. PAHs are not very soluble in water, and therefore tend to adsorb to particles. Sediment from urban rivers can contain 10 to > 100 ppm total PAHs.

Certain PAHs are toxic at environmentally realistic concentrations, causing reduced fertility and immune system suppression in laboratory animals. But the principal concern with PAHs is their carcinogenic potential. PAHs themselves are not carcinogenic, but the metabolites produced by enzymatic metabolism can then covalently bind to DNA. This can result in transcription errors during cell mitosis, which can cause cancer.

PCBs and Dioxins

PCBs have been shown to cause a number of toxic effects in lab animals, including fetal death, tumor promotion, immune system suppression, and liver, stomach, thyroid and kidney damage. Toxic effects in humans have been harder to demonstrate, and include liver damage and a skin disease called chloracne. PCBs are also classified as probable human carcinogens. Like PCBs, dioxins and furans are chemical that are highly toxic. The most toxic and most-studied dioxin is 2,3,7,8-tetrachlorodibenzodioxin

(TCDD). Toxic effects of dioxins and furans in lab animals include chloracne, liver and thymus damage, cancer, birth defects, immune system suppression, and extreme weight loss. The major proven effect in humans is chloracne, though TCDD is listed as a suspected human carcinogen.

Polychlorinated biphenyls (PCBs), dibenzodioxins (dioxins, or PCDDs) and dibenzofurans (furans, or PCDFs) are synthetic chemicals with similar structures, environmental behaviors, and toxicological properties. PCBs were manufactured in the U.S. from 1929 until they were banned by the U.S. EPA in 1977. PCBs had many uses, including electrical capacitors and transformers, plasticizers, and certain paper products. Unlike PCBs, PCDDs and PCCDFs have no practical uses; they are unintended byproducts from certain industrial processes (e.g. paper and pesticide manufacturing), and from waste incineration.

Solvents

Solvents commonly found as environmental contaminants include benzene, toluene, and petroleum-based fuels, and their principal route into the environment is volitilization from industrial processes and leaking storage tanks. The solvents most commonly measured in environmental studies are probably the "BTEX solvents"—benzene, toluene, ethylbenzene, and xylene. These compounds can be measured separately by gas chromatography, or as a group by enzyme immunoassay. The principal human exposure route is via the respiratory system, where solvent vapors diffuse into the bloodstream. A secondary exposure route is absorbtion through the skin, in people who work with solvents. Toxic effects caused by solvents include central nervous system disturbance, liver damage, and blood-related disorders including bone marrow toxicity and leukemia. Many solvents are more toxic after undergoing biotransformation by MFO enzymes.

Pesticides

Pesticides are a group of very diverse chemicals which are unique from a toxicological point of view, in that they are specifically designed to (1) cause toxicity in certain organisms when used as directed, and (2) persist in the environment for a certain length of time. Early pesticides included copper and arsenic compounds and naturally occurring insecticides such as pyrethrum, nicotine, and rotenone. Modern pesticides fall into four groups based on the intended target; insecticides (insects), herbicides (plants), fungicides (molds and other fungi), and others (rodenticides, molluscicides, algicides, etc.). Most pesticides may also be divided into four main chemical groups; organochlorine compounds (e.g., DDT, dieldrin, endrin), organophosphates (parathion, malathion, chlorpyrifos [aka, Dursban], carbamates (Sevin, aldicarb), and pyrethroids. Not counting infrequent, accidental poisonings from occupational exposures (workers in pesticide manufacturing plants, or pesticide applicators), the principal exposure route of pesticides to humans is in the diet.

The toxic effects and mechanism of action for pesticides varies with their chemical structure. Chlorinated pesticides have been linked to liver cancer, birth defects, neurotoxicity, suppressed immune systems, decreased learning ability, and decreased fertility in lab animals, and are known to cause neurotoxicity and genetic damage in humans. Certain organochlorine pesticides are also suspected human carcinogens. The neurotoxic effects of chlorinated pesticides are caused by a disruption of ion balance in sensory and motor nerve fibers, which interferes with the transmission of electrical impulses along the nerves. Pyrethroid pesticides are similar to chlorinated pesticides in effects and mechanism of action, though they're less toxic to mammals because of higher biotransformation and excretion rates. Organophosphate and carbamate pesticides have been shown to cause abdominal cramps, heart disease, muscle weakness, convulsions, coma, and respiratory failure in lab animals, and have caused similar effects in humans accidentally exposed to acute doses. Both of these pesticide groups inhibit the synthesis of the enzyme acetylcholinesterase, resulting in the accumulation of acetylcholine in the central nervous system and the repeated, inappropriate firing of nerve cells.

Pesticides in general may be measured in water, soil, sediment, or biological tissues by gas chromatography, or more easily and cheaply by enzyme immunoassay.

Assessment Using a Bioassay

A foundation of toxic assessment is the concentration-response relationship. As concentrations of a chemical or effluent increase, the response as observed in a test organism should become more severe. Toxicity of a substance is a function of the concentration of a substance and the duration of exposure. If concentrations of a toxicant or effluent are high and the duration of exposure is low, this often results in acute effects. When concentrations of a toxicant are low, but the duration of exposure is prolonged, this often results in chronic effects. Effects as measured in bioassays are referred to as endpoints. Endpoints may be death of an organism in acute bioassays. In the lettuce seed bioassay, there are two endpoints: seed germination rate and root length growth.

Advantages of Bioassays

➤ Integrative—measures the effects of all chemicals in a sample.

➤ Embraces sampling design, experimental design, hypothesis testing, interpretation of data, and the disciplines of biology, chemistry, and mathematics.

➤ Serves as a platform for discussing risk assessment and the pathways and fates of chemicals in the environment.

Limitations of Bioassays

➤ Methods not as standardized as chemical analyses.

➤ Does not identify what is causing the toxic response/effect.

➤ Usually tests only one species at a time; cannot extrapolate results to the community level.

Lettuce Seed Bioassay

The idea of using plant seeds in bioassays originated from studies of herbicide toxicity for agriculturally important crops. These assays have been recommended by the U.S. EPA, the U.S. Food and Drug Administration, the U.S. Department of Agriculture, and the Organization for Economic Cooperation and Development. In general, this bioassay (and most plant bioassays) is more sensitive to herbicides than animal bioassays, about equally sensitive to metals, and less sensitive to some insecticides and industrial chemicals.

This bioassay can reveal the relative toxicity of point sources of pollution, or of a single discharge over time. A geometric series of test concentrations of sample water may be made by multiplying the highest and succeeding concentrations by a constant factor (0.6 to 0.2). For example: 100 percent effluent, 50 percent, 25 percent, 12.5 percent, and 6.25 percent. See the guideline below. The dilution water, should be the same as the control water. A common design is to use water from a suspected point source as the "test water," and grab samples of relatively uncontaminated river water (upstream of the outfall) as the "dilution water."

Test Concentration	Test Water (mL)	Dilution Water (mL)
Control cups	0.00	100.00
100 percent effluent	100.00	0.00
50 percent effluent	50.00	50.00
25 percent effluent	25.00	75.00
12.5 percent effluent	12.50	87.50
6.25 percent effluent	6.25	93.75

Safety

The overriding concern in collecting samples from effluents, other point sources, and sediments is potential exposure to hazardous levels of toxicants. With this in mind, the following safety guidelines are offered.

Contact the nearest Health Department, Environmental Protection Agency, or Department of Natural Resources or Environment to get a list of toxic contamination sites, and a ranking of the severity of contamination. Some river reaches, sediments, or point sources may contain dangerously high levels of toxicants.

The chemical composition of most effluents are largely unknown; every effluent should be considered a potential health risk and exposure should be limited.

When sampling always wear rubber gloves.

Wash exposed parts of the body with soap and clean water immediately after collecting samples.

Always sample with another person.

Containers used to hold collected water samples should be clearly marked.

Materials

➤ Control water, distilled or dechlorinated tap water

➤ Chlorox: 10% solution, in control water

➤ Petri dishes (9 or 10 cm diameter)

➤ Whatman #1 filter paper disks (9 cm diameter)

➤ Pipette capable of delivering 5 mLs

➤ Calculator

➤ Graph paper

➤ Dark location to incubate seeds (inside a desk drawer or cabinet is adequate)

Procedure

1. Soak seeds in a 10% solution of Chlorox for 20 minutes, and rinse seeds 5 times with control water (this step kills fungi that may interfere with seed germination).

2. Place filter paper into a petri dish. Label dish with appropriate single number (at least three dishes per sample are recommended, for statistical purposes).

3. Pipette 5–7 mL of sample water onto the filter paper; enough to saturate the paper (please use a consistent volume in all the tests).

4. Place 10 seeds onto the filter paper.

5. Also prepare dishes for the control sample: 5–7 mL of control water, 10 seeds, etc.

6. Incubate the dishes in the dark at room temperature for 5 days (120 hours). Temperature should be fairly stable during incubation and should not exceed 85°F.

For each dish, record:

➤ seed germination rate (number of seeds that germinated (10 x 100%))

➤ individual root lengths for each germinated seed (to the nearest mm)

➤ if replicate samples were analyzed, calculate the mean and standard deviation of the seed germination rates and root lengths for each sample and the control;

➤ repeat the test if the germination rate for the control sample is less than 80%; something is wrong with the control water, the seeds, or the incubation conditions.

Notes

While it is always best to use fresh seeds in this bioassay, most seeds will remain viable for up to one year if stored in a refrigerator.

It is OK to briefly check on the dishes during the incubation period. If the paper has dried, pipette a few mLs of control water onto the paper. Do not use additional sample water; this could add more contaminants.

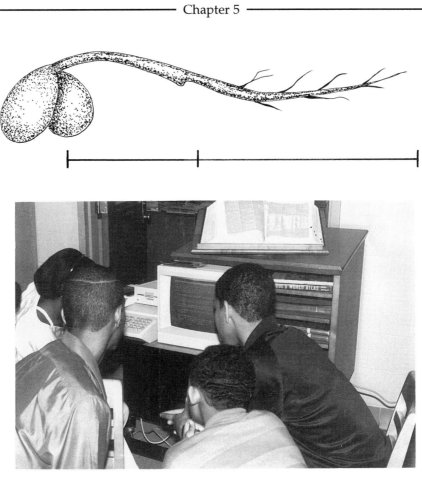

Figure 5.1. Students using Internet to learn more about the impact of toxics on the aquatic environment.

Benthic Macroinvertebrates

Characteristics of Benthic Macroinvertebrates

Have you ever wondered what lives in a stream or river? Fish, of course, is a natural answer, but there is more. A river is filled with life, from tiny microscopic organisms that can be seen with the unaided eye, to insects that cling to rocks and burrow into silt.

The term "benthic" means bottom dwelling and refers to organisms that live in, crawl upon, or attach themselves to the bottom (substrate). The term "macroinvertebrates" refers to those invertebrates seen with the unaided eye. Most benthic macroinvertebrates in flowing water are aquatic insects or the aquatic stages of insects, such as stonefly nymphs, mayfly nymphs, caddisfly larvae, dragonfly nymphs, and midge larvae. They also include such things as clams and worms. Examples of some of these creatures are shown on the following pages.

Benthic macroinvertebrates often go unnoticed because of their size and habitat, but they are an extremely important part of river ecosystems. Collecting benthic macroinvertebrates can provide a greater understanding of a river's condition. Several characteristics of benthic organisms make them useful indicators of water quality:

1. many are sensitive to physical and chemical changes in their habitat;

2. many live in the water over a year;

3. they cannot easily escape pollution as some fish can; and

4. they are easily collected in many streams and rivers.

Life Cycles

Aquatic insects go through several stages from egg to adult. The number of stages depends on the type of metamorphosis (series of developmental changes) followed. Insects that undergo *incomplete metamorphosis* follow three stages. They begin as eggs that hatch into nymphs, which then grow into adults. The immature period is called the nymphal stage. Examples of insects that undergo incomplete metamorphosis are mayflies, dragonflies, stoneflies, and true bugs.

Many of the insects that undergo incomplete metamorphosis are aquatic only during the egg and nymphal stages. The winged adults do not live in the water. Dragonflies, for example, can often be seen in the adult form flying along streams and rivers in the summer.

In contrast, true bugs begin life in the water and grow to become aquatic adults. Examples include backswimmers, water scorpions, and water striders. Each has a different strategy for obtaining oxygen. Backswimmers surface for air, water scorpions obtain oxygen through a breathing tube (like a snorkel), and water striders skim along the water surface.

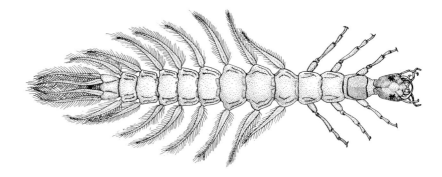

ACTUAL SIZE

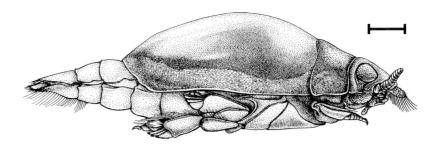

Figure 6.1. Larva and adult of a whirligig beetle. Both forms are aquatic. (Drawing by A. Holt.)

Insects that undergo *complete metamorphosis* have four stages: egg, larva, pupa, and adult. They begin as eggs that hatch into tiny larvae; these larvae grow and eventually enter a pupal stage in which the insects are transformed into an adult. The immature period is called the larval stage. Examples of insects that go through complete metamorphosis are true flies, beetles, and caddisflies.

Again, many insects that undergo complete metamorphosis are aquatic during the egg, larval, and pupal stages, but not as adults. On the other hand, some insects pupate out of the water in overhanging tree branches (like the whirligig beetle and predaceous diving beetle), but the adult form is aquatic.

Aquatic insect life cycles range from less than two weeks for some midges and mosquitoes, to two years or longer for some stoneflies, dragonflies and dobsonflies.

Note that some benthic macroinvertebrates described in this chapter are not insects: leeches (Hirudinea), aquatic worms (Oligochaeta), water mites (Arachnida), snails (Gastropoda), and clams (Pelecypoda).

Major Benthic Groups

The following is a brief description (adapted from River Watch Network) of the major taxonomic groups of benthic macroinvertebrates that may be found in a stream or river.

Caddisflies

The caddisflies (Order Trichoptera) make up a large part of benthic communities. Some species are free-living while others make case retreats out of silk, sand grains, pebbles, or bits of plant matter (see Fig. 6.3).

All caddisflies have hard-shelled head capsules. Sometimes the first three segments behind the head also have hard-shelled plates on the top (dorsal) surface above the attachments for three pair of legs. The rest of the body (abdominal segments) is soft and often cylindrical. The larvae possess two small hooks on the last segment.

Caddisflies undergo complete metamorphosis and the larvae transform into winged adults in the water. As adults, caddisflies only live a few days and do not eat at all.

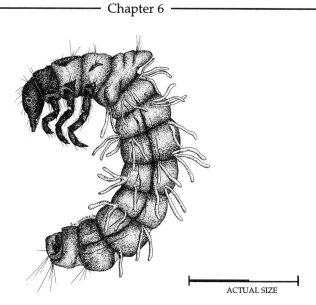

Figure 6.2. Caddisfly larva. (Drawing by S. W. Downes, a 14-year-old student in a water quality monitoring project.)

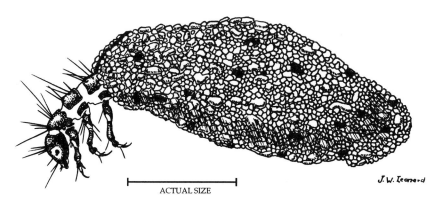

Figure 6.3. Caddisfly larva and its case made of small stones. (Drawing by J. W. Leonard.)

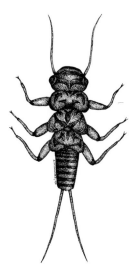

ACTUAL SIZE

Figure 6.4. Stonefly nymph. (Drawing by
S. W. Downes.)

Stoneflies

Stoneflies (Order Plecoptera) are indicators of good water quality
because the nymphs require highly oxygenated water. They tend to inhabit
clear cold streams, and are highly intolerant of changes in water quality. A
stonefly nymph is shown in Figure 6.4.

Stoneflies undergo incomplete metamorphosis. The aquatic nymphs
transform directly into winged adults. The heads and top surface of the first
three body segments on nymphs are hardened. Their antennae are moderately long to long, and all species have exactly two tail filaments (or cerci).

Small stonefly nymphs can easily be mistaken for mayfly nymphs;
however, mayfly nymphs usually have three tail filaments and stonefly
nymphs have gills around the base of their legs or no gills at all.

Mayflies

Mayflies (Order Ephemeroptera) are usually easy to identify. The
nymphs can be small and squat, or long and slender. They have three pairs
of segmented legs and visible antennae. They are most easily identified by
their three (rarely two) tail filaments, and by the seven pairs of abdominal
gills found on most species. The gills may be either flat and spade-shaped,
or feathery in appearance.

Mayfly nymphs are often flattened or streamlined to reduce the force
of fast currents. They are most abundant in clear streams, though a few
kinds may be found in other habitats. See Figure 6.5.

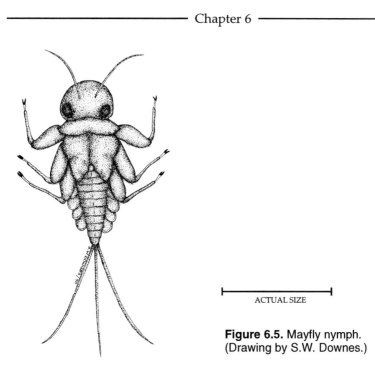

ACTUAL SIZE

Figure 6.5. Mayfly nymph.
(Drawing by S.W. Downes.)

Alderflies, Dobsonflies, Fishflies

These insects (Order Megaloptera) include some of the largest aquatic larvae (1.5-3 inches). There are two families known from North America. The larvae are most often found in clean rivers and streams with rocky bottoms.

The head capsule and the first three body segments of the larvae are hardened. They are recognizable by the presence of lateral filaments extending out from the sides of each abdominal segment. The larvae are very mobile and predaceous and have been known to pinch hands. A dobsonfly larvae known as a hellgrammite is shown in Figure 6.6.

Midges

The midge family (Family Chironomidae, Order Diptera) is one of the largest in North America, with over 2,500 species. They may be quite abundant, particularly in polluted streams and rivers. Some midges have hemoglobin, allowing them to survive in habitats with little dissolved oxygen.

Midges have small larvae, usually around 1/4" in length. The larvae lack jointed legs, as do the larvae of other "true flies." However, they have a pair of small "prolegs" just below the head, and another pair posteriorly (see Figure 6.7). The head is hard while the rest of the body is soft.

Midge larvae vary in color; they may be white, cream, tan, pale red, or pale green. They differ from other fly larvae in that they are small, slender, and almost always curved into a "C" or "S" shape.

ACTUAL SIZE

Figure 6.6. Dobsonfly larva (hellgrammite).

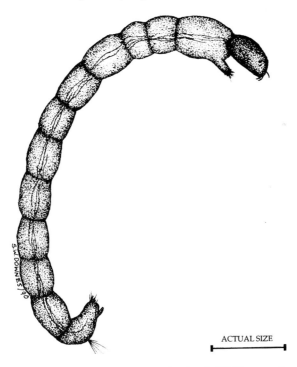

ACTUAL SIZE

Figure 6.7. Midge larva. (Drawing by S. W. Downes.)

Aquatic worms may be mistaken for midge larvae, but worms have many segmented, limp bodies and back legs or other external features.

Craneflies

Craneflies (Family Tipulidae, Order Diptera) make up another of the largest family of flies. The larvae tend to be oblong, cylindrical, and somewhat tapered toward the head. The head is retractable and only partially hardened. The last abdominal segment usually has several finger-like lobes.

Other True Flies

The other true flies (Order Diptera) include many families with aquatic or semi-aquatic larvae, such as blackflies, horseflies, sandflies, no-see-ums, deerflies, and sewageflies. These vary in size and shape, but all lack jointed legs. Some have complete, exposed head capsules; others have a reduced, retracted head. The bodies are soft and flexible. A blackfly larva is shown in Figure 6.9.

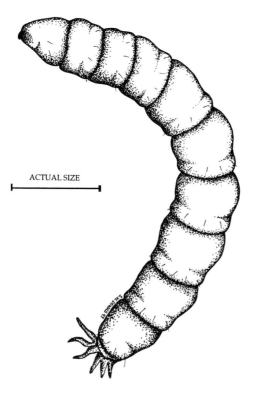

ACTUAL SIZE

Figure 6.8. Cranefly larva. (Drawing by S. W. Downes.)

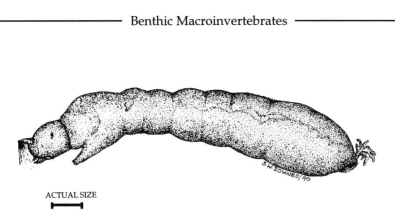

ACTUAL SIZE

Figure 6.9. Blackfly larva. (Drawing by S. W. Downes.)

Dragonflies and Damselflies

Dragonflies and damselflies (Order Odonata) are predators. Their pre-historic-looking nymphs (Figures 6.10 and 6.11) are not common in fast-moving streams, but may be abundant in sluggish waters. They may be elongated, and are a somber grey, green, or brown color. The body may be smooth or rough and is often covered with a growth of algae and organic debris.

Water Mites

Water mites (Hydracarina) are small aquatic relatives of spiders. They are nearly round or globular, and have eight legs. They are not likely to be confused with any other aquatic organism. One of the common water mites is bright red and not much larger than the periods on this page.

Snails

Snails (Order Gastropoda) are easy to identify. Most have a spiral shell, though a few have a shell in the shape of a cone (limpets). Snails with different shaped shells usually are from different families. An important character in family identification is whether the shell spirals to the left or right.

Aquatic Worms and Leeches

Aquatic worms (Order Oligocheata) have the same multisegmented appearance as the common terrestrial earthworms. Leeches (Order Hirundinea) are many-segmented but appear flattened and have a sucker on both ends of the body (see Figure 6.12).

Neither group has head capsules or any other obvious physical distinctions. There are ten families of aquatic oligocheats; the majority of these occur in areas high in organic debris or mud. For this reason, they are usually considered a tolerant group for both siltation and organic pollution.

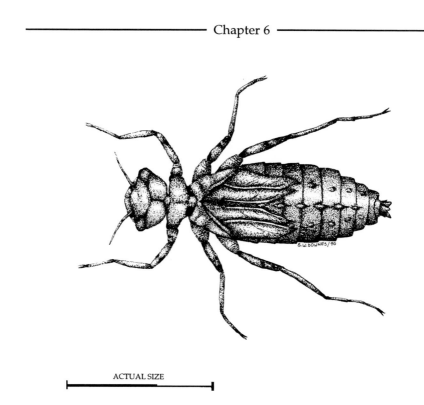

ACTUAL SIZE

Figure 6.10. Dragonfly nymph. (Drawing by S. W. Downes.)

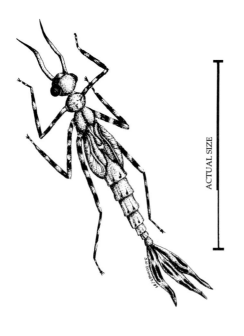

ACTUAL SIZE

Figure 6.11. Damselfly nymph.
(Drawing by S. W. Downes.)

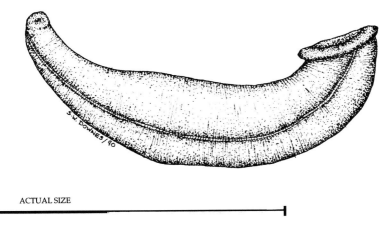

ACTUAL SIZE

Figure 6.12. Leech. (Drawing by S. W. Downes.)

Aquatic Sowbugs

Aquatic sowbugs (Order Asellus) are omnivorous. They look very similar to their terrestrial relations known by some people as roly-polys. They are microcrustaceans and are often abundant on rocks in slower-moving waters. These organisms are usually dull-gray to black in appearance and are quickly identified by the many appendages lining each side of the body (see Figure 6.13).

Functional Feeding Groups

Aquatic insects can be grouped according to how and what they eat. These different "functional feeding" groups, described below, reveal the many roles aquatic insects play in stream ecosystems.

Shredders feed on coarse dead plant material: leaves, grasses, algae, and rooted aquatic plants. Shredders include stonefly nymphs, caddisfly larvae, and some fly larvae like the cranefly.

Shredders break large pieces of organic matter into finer material that is released in their feces. Although they appear to be herbivores (plant-eaters) shredders and other functional groups are really omnivores because they consume living organisms along with dead plant material. This happens because dead plant material in water is rapidly colonized by bacteria, fungi, and protozoans (one-celled organisms). It is thought that insect shredders obtain some of their nutrition by eating these attached organisms.

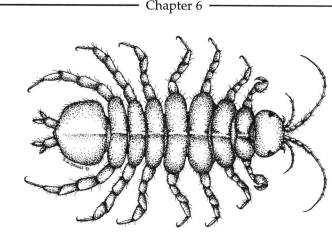

ACTUAL SIZE

Figure 6.13. Aquatic sowbug. (Drawing by S. W. Downes.)

Collectors feed on decomposing organic matter that includes the feces of upstream organisms and plant fragments. Bacteria attached to this material become a source of nutrition for collectors. There are two types of collectors: those that filter and those that gather.

Filtering collectors strain minute particles out of the water. The black fly larva, for example, is found in fast current areas where it catches small organic particles with its extended fan-shaped brushes. Net-building caddisflies are also included in this group.

Gathering collectors feed mostly on dead organic material on the river bottom. This group is represented by mayfly nymphs, caddisfly larvae, adult beetles, and fly larvae like the midge. Now that you know what midge larvae eat, you can understand why they are most often found in bottom sediments where dead organic material accumulates.

Scrapers graze on periphyton (attached algae) growing on stones and other substrates in the water, or on surface vegetation. Scrapers are usually exposed to the current and so have developed morphological and behavioral adaptations, such as flattening of the body, for maintaining their position. Scrapers include many mayfly nymphs, caddisfly larvae, and water penny or riffle beetle larvae (Figure 6.14).

Predators feed on other aquatic insects and have adaptations specifically for the capture of live prey, such as scoop-like lower jaws (dragonflies and damselflies), large pincer-like jaws (hellgrammites), and spear-like mouth parts (water striders). Other predators include large stonefly nymphs, true bugs (such as water striders), some caddisfly larvae, and some beetle larvae and adults.

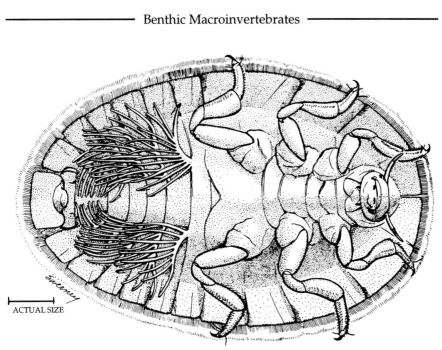

ACTUAL SIZE

Figure 6.14. The flattened shape of a water penny, a representative scraper, that allows it to live in the boundary layer. (Drawing by K. Sweeney.)

Physical Forces that Influence the Benthic Community

The distribution and relative abundance of aquatic insects in a stream or river are the result of many factors. The most significant of these are:

1. Water temperature patterns
2. Discharge patterns (volume and velocity of flow)
3. Substrates (stream bottom composition)
4. Energy (trophic) relationships

As you travel along a river from its headwaters to its mouth, you will notice obvious differences in water temperature, increases in water volume, changes in the river bottom, and variations in the sources and quality of food (energy relationships). Some of these changes are related to the type and use of the land the river flows through; some are a function of increasing stream order.

Let's investigate these environmental factors and the physical and behavioral adaptations of benthic macroinvertebrates to them.

Figure 6.15. Elementary student noting diversity of aquatic life in the riffle area.

Water Temperature

Daily and seasonal patterns of water temperature affect macroinvertebrate metabolism, growth, emergence, and reproduction. Metabolism of aquatic insects includes the use of food energy, feeding, and growth rates.

Warmer temperatures usually cause an increase in the growth rate of aquatic organisms, while lower than normal temperatures decrease growth rates. Many aquatic organisms are active during winter when water temperatures hover around 2°C (36°F). Some insects, such as stonefly nymph, are adapted to cold water and do best under these conditions.

Figure 6.16. A high school class studying river ecology and adaptations of aquatic organisms to living in flowing water.

The life cycles of many aquatic insects are closely tied to seasonal water temperatures. For example, water temperature influences the onset of emergence—the time when aquatic insects become winged adults. The period of emergence varies with water temperature for many species, but in general, warmer than normal temperatures lead to earlier emergence.

Similarly, for many aquatic insects, eggs will only hatch in response to "thermal cues." Other insects have life cycles that are synchronized to take advantage of favorable periods, such as the fall leaf drop for shredder nymphs and larvae. Some aquatic insects are also capable of diapause—a period of reduced metabolism to avoid unusually warm or cool water temperatures, as well as other unfavorable conditions.

Temperature patterns vary along a river from its source to mouth. In the upper reaches of a river, headwater streams are often fed by groundwater that bubbles out at a constant cool temperature. In forested areas, these streams may be entirely shaded by overhanging tree branches. As a result, they may experience only small daily or seasonal changes in water temperature. Such streams provide good habitat for organisms adapted to cooler water, such as stonefly nymphs.

On the other hand, headwater streams in prairies or grasslands, where trees are scarce, or streams fed largely by runoff from the land, tend to show greater seasonal and daily changes in water temperature. Sunlight and air temperature have a greater impact on water temperature in these streams.

In a river's midreach, the river widens to the point that trees can no longer shade the entire water surface. The sun is able to warm the water, causing a slight increase in water temperature during the day (accompanied, of course, by a slight drop in water temperature at night). The current also slows in the midreach, allowing more absorption of the sun's rays and further warming of the water. It is in this section of the river that seasonal changes in water temperature are the greatest.

Rivers generally carry more water in their lower reaches. The larger volume of water prevents significant daily changes in water temperature, since much more energy is required to warm a large mass of water than a smaller mass by the same amount. As you might expect, water temperatures in the lower reach of a river generally fluctuate less than in the midreach.

Discharge Patterns

The current or flow of water is common to all rivers. Current sets rivers and streams apart from lakes and ponds, and is a major factor in shaping river ecosystems. Aquatic plants and animals depend upon a river's current to bring food and nutrients from upstream, and to flush wastes downstream.

Discharge is a measure of the volume of water passing a given point over a given period of time. This measure is often expressed as cubic feet or meters per second.

Measuring velocity and calculating discharge can give much information about a river system. A method for measuring and calculating discharge is described later in this chapter.

The speed of a river's current is controlled by four factors:

1. Depth
2. Slope or steepness of the land
3. Width of the stream channel
4. Roughness of the river bottom

The current is fastest just beneath the water surface because friction is decreased between the air and surface water. As water becomes deeper the velocity is reduced, although the volume of water is increased. Velocity also depends on the type of substrate—gravel and rock bottoms provide more friction to flow than sand or silt.

A river's discharge depends upon a variety of factors. One of these is the nature of the surrounding lands. If the land can absorb and retain rainfall (for example, where soils are sandy or gravel), changes in discharge are usually small. On the other hand, where the land is less able to absorb rainfall, as in urban areas covered by concrete or asphalt, changes in discharge after each rainfall will be relatively great.

Stream discharge may also show seasonal increases during the spring snowmelt and during rainy seasons.

Benthic macroinvertebrates have evolved to meet the challenges of living in a river current. During floods, many macroinvertebrates burrow into the substrate to avoid being swept away. Although floods do "scour" the riverbed by shearing off attached algal communities and carrying organisms downstream, they are natural occurrences that serve to disperse nutrients and reintroduce organisms from upstream.

As water flows over rocks, a protective boundary layer about 1-4 mm thick is formed near the rock's surface in which the current drops to near zero. Flattening of the body has enabled macroinvertebrates such as the stonefly nymph and water penny to "crouch" into the boundary layer. Flattening also permits organisms to avoid the current by crawling under rocks.

Some organisms have evolved elaborate means of maintaining their position in the current. The blackfly larva (Simulium), for example, has a ring of many tiny hooks at its enlarged hind end. The larva also has salivary glands that produce silk; this silk is woven into a mat on a fixed object in the riverbed. The larva grasps the silk by its rear hooks freeing the head to spin another silken mat. If the current becomes too swift, the larva draws itself into the quiet boundary layer. A silken thread is attached to the river bottom serving as a safety line in case the larva is swept from its mat. If this occurs. The blackfly larva can crawl along this silken strand to its original position.

Many caddisfly larvae construct homes (cases) of stones, vegetation, sand particles, or other river bottom material (Fig. 6.3). The primary purpose of these retreats is to provide protection against predators, but they can also act as a weight or ballast to prevent the larva from being carried downstream. Caddisfly larvae also possess claws on their rear appendages to hook onto their cases.

Substrates

The type of river bottom or substrate—bedrock, rocks, gravel, sand, or silt—is determined in part by the current velocity and by the underlying geology of the area. Rocks, small stones (rubble), and bedrock are characteristic of substrates in fast flowing streams. Rocks and rubble areas offer numerous niches for benthic macroinvertebrates. Fast river currents can carry particles of 1/4 inch in diameter or less downstream, leaving only larger rocks and rubble in place.

There are also substrates not associated with the stream or river bottom, such as leaf packs that accumulate along the upstream edges of fallen logs and branches. Attached algae and rooted aquatic plants are considered living substrates.

Aquatic organisms have developed specific adaptations for living on or in a particular substrate. For example, the substrate is the source of building materials used to construct caddisfly larva cases. Some caddisfly larvae like *Hydropsyche* depend upon small stones to anchor their tiny silken nets. Stonefly nymphs possess claws that allow them to sprawl on the bottom hooked to small stones.

As rivers become larger, slower, and deeper, the river bottom becomes silty from accumulated upstream sediments. In slower river reaches, less mixing between the river and atmospheric oxygen takes place, leading to lower dissolved oxygen levels in bottom sediments. Organic pollution often decreases the dissolved oxygen level in the sediments even further.

Organisms that live in silt and sand bottoms have special adaptations for life there. Some species of the midge larva *(Chironomus)* are good examples of pollution-tolerant organisms well adapted to silty river bottoms. They possess a form of hemoglobin—the source of this organism's red color—that enables them to utilize the small amounts of dissolved oxygen available in these habitats.

The sewage worm *(Tubifex)* exists very well in soft substrates like silt. This organism is a segmented worm similar to its relative the terrestrial earthworm. While it feeds with its mouth buried in the bottom, its tail waves back and forth in the water, creating a current that brings it fresh supplies of dissolved oxygen.

Another benthic organism adapted to silty river bottoms is the mayfly nymph *(Hexagenia)*. Using its heavy forelegs, the nymph burrows a "U-shaped" tunnel in the soft substrate, open at both ends. The mayfly creates a water current within its hole by rhythmically beating its large feathery gills, bringing it a fresh supply of oxygen as well as food. Its tunnel also helps protect the nymph from predators. However, unlike midge larvae and sewage worms, *Hexagenia* is not tolerant of very low dissolved oxygen levels.

Energy (Trophic) Relationships

The quantity and quality of food available to aquatic organisms strongly affects the kinds and numbers of organisms found in a river.

The headwaters of a river system are very important to the health of the entire river, because this is the source of much food carried downriver. In forested areas, headwater streams are cloaked by overhanging trees and shrubs that provide much of the food energy available to aquatic organisms. Energy enters the stream in the form of leaves, fruits, nuts, berries, twigs and bark. Some of this material becomes fine material, but most is available as coarse material. Because of this, collectors and shredders are both common in forested headwater reaches.

In prairie, mountainous, and desert regions, headwater streams may not be shaded by streambank vegetation. Attached algae and rooted aquatic plants are relatively more important sources of food energy than streambank vegetation in these streams. As a result, grazers are more common than shredders.

In a river's midreach, flow increases and the stream becomes wider. Riparian vegetation is no longer able to cover the whole stream, and sunlight reaches the water. Here periphytic growth (attached algae) becomes more abundant, and aquatic grazers and collectors are common. Organisms like mayfly nymphs shear off pieces of attached algae growing on rocks. Collectors feed upon transported fine material (shredder feces and reduced organic pieces) from the headwaters and from local vegetation.

As a river widens and gathers more flow, it becomes deeper and more turbid, thereby limiting sunlight penetration. Rooted vascular plants may grown along the shoreline and some attached algae may also grow in the shallows if stones or other substrate are available. Collector organisms predominate in this river reach, filtering out minute particles suspended in the water and gathering fine particles that have settled to the river bottom.

An activity designed to look at the relationship between shredders and coarse materials is described later in this chapter.

Benthic Macroinvertebrates as Indicators of Water Quality

The relationship between the composition of the aquatic community and water quality has long been recognized. Two commonly used methods for evaluating water quality by looking at macroinvertebrates are indicator organisms and diversity indices.

The concept of indictor organisms is based on the fact that every species has a certain range of physical and chemical conditions in which it can survive. Some organisms can survive in a wide range of conditions and are more "tolerant" of pollution. Others are very sensitive to changes in conditions and are "intolerant" of pollution. Some examples of pollution-sensitive organisms are mayflies, stoneflies, some caddisflies, riffle beetles, and hellgrammites. Examples of pollution-tolerant organisms are sludge worms, leeches, and certain midge larva.

The evaluation of water quality is linked to the numbers of pollution-tolerant organisms at the site compared with intolerant organisms. A large number of tolerant organisms and few or no intolerant organisms would indicate pollution. (Note, however, that pollution tolerant organisms can be found in both polluted and unpolluted waters.)

Figure 6.17. High quality rivers support a diverse population of aquatic organisms, like this riffle area.

Diversity refers to the number of different kinds of organisms found in a biological community. Greater diversity means more different types of organisms. In general, communities with high diversity tend to be more stable than those with less diversity.

Pollution tends to reduce the number of species in a community by eliminating organisms that are sensitive to changes in water quality. The total number of organisms may not change, however, if pollution-tolerant organisms increase in abundance.

Qualitative and Quantitative Studies

Qualitative studies are used to gather baseline data. Collecting is done from as many types of macrohabitats at a site as possible to get an overall impression of the kinds of organisms present in the stream. The information gathered is useful for comparisons over time at the site but evaluation of the community is not necessarily made.

Quantitative studies are used to evaluate the macroinvertebrate community at the sampling site and to make comparisons between sites. In quantitative studies, it is important to match the physical characteristics of the sites being evaluated, and to expend the same amount of effort in sampling. The goal is to control all variables between sites except water quality so that any differences in results can be attributed to water quality.

Admittedly, it is very difficult to sample a subject as dynamic and varied as a river in a truly quantitative manner. Useful comparisons can be made, however, by matching sampling time, methods and site conditions as closely as possible, or by using artificial substrates.

Sampling Methods and Equipment

There is a variety of equipment and methods for sampling benthic macroinvertebrates; the choice of which to use depends on the character of the river, the site to be sampled, and the preference of the persons doing the sampling.

With natural substrate sampling, organisms are collected from the existing stream bottom. Only one trip to the site is needed, and sampling is done from a riffle area, i.e. a somewhat shallow area with a steady current and rock and gravel substrate.

Artificial substrates are often used when it is not possible to sample the actual stream bottom, or when sampling needs to be more standardized. Artificial substrates, like the Hester-Dendy sampler (Fig. 6.21), provide a site for organisms to colonize. Such samplers must be kept in the water for 4-6 weeks to allow for colonization.

Qualitative Sampling Devices

Some commonly used sampling devices are described below. They are generally considered qualitative samplers because the area sampled is not precisely measured. The standard netting size used in these samplers is U.S. #30 mesh—or 0.589 mm spaces—as this is the size used to differentiate organisms visible and not visible to the unaided eye.

D-Frame Nets

D-Frame nets are a basic tool for collecting benthic macroinvertebrates (Fig. 6.18). They work well because the flat area can rest on the bottom, preventing the loss of organisms underneath the net. They can also be used for sampling along banks and in vegetation.

D-Frame nets are relatively easy to obtain, but they can also be easily constructed with an old pole or stick, a hanger, and some netting.

Kick Screens

Kick screens work well in sampling riffle areas, and they permit a larger area to be sampled than with D-Frame nets. The kick screen consists of screening material stretched between two poles (Fig. 6.19). Sampling is done by pushing the two poles into the substrate until the edge of the screen rests on the bottom. Organisms are dislodged by disturbing the substrate on the

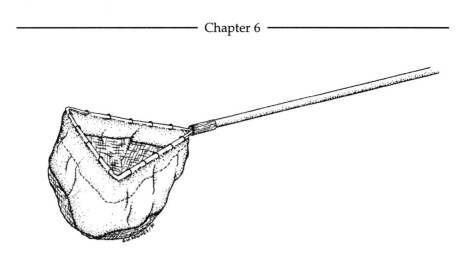

Figure 6.18. D-frame net. (Drawing by S. W. Downes.)

upstream side of the stream, allowing them to be carried by the current into the screen. Kick screens can be easily constructed from window screening and two wooden poles.

Kick Seine
A kick seine is similar to the kick screen, except that the netting is generally nylon mesh. Also, a seine usually has weights along the bottom and floats along the top to help keep it perpendicular to the current. Sampling is basically done the same as with a kick screen.

A seine may be constructed from two wooden poles and nylon mesh (Fig. 6.19) or a burlap sack.

Quantitative Sampling Devices

Quantitative samplers are used to sample a set amount of bottom area. They are more consistent from sample to sample then hand nets and seines. Many of these samplers are also developed for areas that can't be reached with nets and seines.

Surber or Square-Foot Sampler
In this type of sampler the net is held open by a square foot frame hinged to another frame of the same size (Fig. 6.20). The cloth sides on the frame prevent eddying and help keep the current flowing into the net. During sampling, the frame is pushed into the substrate and the area within the frame is disturbed by rubbing rocks. Dislodged organisms are carried by the current into the net.

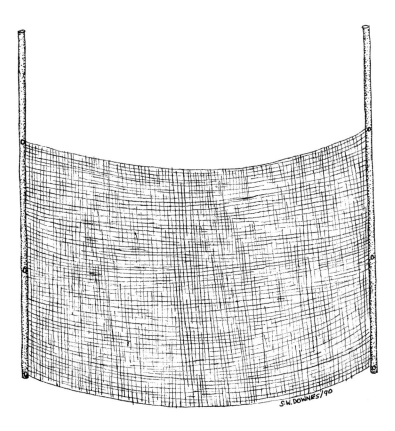

Figure 6.19. A kick screen used for sampling shallow river reaches. (Drawing by S. W. Downes.)

Dredge or Grab samplers

These samplers are generally lowered from a bridge or boat to the substrate, and are tension or spring-released. There are different types for a variety of substrates and situations. Two of the most widely used dredge samplers are the Peterson and the Ekman.

The Peterson can be used in sand, gravel, marl, or clay; it is generally used in deep water and swift currents. The Ekman is used in silt, muck, or sludge; it is used in water with little current. More information on Dredge and Grab samplers can be found in the *Standard Methods for the Examination of Water and Wastewater.*

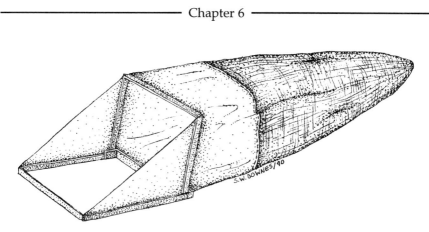

Figure 6.20. A Surber or square-foot sampler. (Drawing by S. W. Downes.)

Artificial Substrate Samplers

Artificial substrates can be useful in situations where it is not possible to sample the natural substrate, such as in large rivers, channelized areas, or deep water. They can also provide standardized sampling as long as setting and retrieving procedures are the same for each site. The drawbacks of artificial substrates are that they may be difficult to retrieve without disturbing organisms and allowing them to escape. Vandalism is sometimes another problem.

The artificial substrates should be placed in the river for 4–6 weeks to allow for colonization. It should be kept in mind that the artificial substrate may select for certain types of organisms; as a result, the sample will not fully represent the diversity of organisms at the site. However, the information gathered is still useful for learning about the river ecosystem.

Hester-Dendy or Multi-plate Sampler

This sampler is made of stacked rough-textured square plates separated by smaller plates or washers, and is suspended in the stream (Fig. 6.21). The large plates should be about 3" square (7.6 cm) and 1/8" thick (0.3 cm). The smaller plates are usually about 1" square (2.5 cm) and 1/4" thick (0.6 cm). The plates can be made of hard board, cement, unglazed porcelain or anything rough and durable. They need to have holes in the middle so they can be threaded together.

A Hester-Dendy sampler can be easily constructed with eight large plates, seven small plates or washers, and one eyebolt or threaded rod and two nuts or nylon cording to hold the plates together.

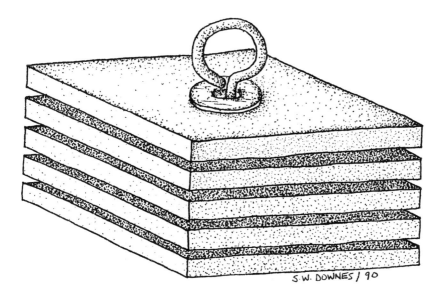

Figure 6.21. Hester-Dendy sampler, a type of artificial substrate. (Drawing by S. W. Downes.)

Reuven Ortal River Sampler

This sampler can be constructed from 1/4" fiberglass (6 x 6 mm) or plastic mesh (see Figure 6.22). The 7" x 7" flat sampler (18 x 18 cm) can be sewn together on three sides and then filled with the same river bottom sediment that the sampler will be placed on. Then sew up the remaining side and place the sampler on the bottom of the river to be colonized by macroinvertebrates. A line from the sampler can be attached to a stake on the bank of the river for retrieval.

This kind of sampler might be used in a river that is polluted, and by using gloves could prevent direct contact with the polluted water.

Basket Sampler

This is a wire mesh basket 7 in. (18 cm) in diameter and 1 1 in. (28 cm) long, filled with 30 rocks of 2-3 inches (5-7.5 cm) diameter. The method was developed for collecting macroinvertebrates in large rivers and lakes. The sampler should be suspended at a depth of about 5 ft., within the zone where light penetration reaches and permits the growth of green plants (euphotic zone). The basket sampler can also be placed on a river bottom.

A basket sampler can be easily made from 1/4" (6 mm) hardware cloth, or a commercially available barbecue basket can be used.

I **Figure 6.22.** Reuven Ortal sampler for collecting macroinvertebrates.

Sampling Methods and Equipment

Safety Precautions

a. Be sure to protect yourself from the water. Use waders, gloves, plastic bags, or whatever else is available.

b. When you are finished sampling, wash your hands and any other part of your body that may have come in contact with the water.

c. If you have any open cuts, they MUST be kept out of contact with the water.

d. Do not sample during periods of high water. Not only is the current usually strong and dangerous during these times, but the high flow disrupts the benthic fauna, making sampling unproductive.

e. Beware of sampling in cold weather and in cold water, especially when it is windy. These conditions could lead to hypothermia.

Natural Substrate Sampling Method

The following basic sampling techniques should be followed whether a net, kick screen, seine, or surber sampler is used.

a. Approach your sampling site from downstream to avoid disrupting the organisms at the sampling location before you are ready to collect.

b. Place the net at the downstream side of the sampling area with the opening facing upstream. Hold the net perpendicular to the flow, but at a slight downstream angle if you are using the kick screen or seine. Do not step into the sampling area yet.

c. If you have gloves and the water is clean enough, pick up each stone 2" (5 cm) or more in diameter and rub it to remove any organisms from its surface. Be sure you are holding it under the water and in front of the net. Place each rubbed stone outside of the sampling area.

d. When all stones 2" (5 cm) or larger have been removed, or as many as possible, step into the upstream edge of the sampling area and kick the stream bed vigorously, working your way towards the net. Be sure to use the same amount of time at each site. (Three minutes is sufficient.)

e. Remove the net with an upstream scooping motion. If using a kick screen or seine, do not allow water to flow over the top. If you are using a net, you may want to carefully drag the net through the water to work the sample to a bottom corner so it will be easier to remove.

f. Bring the net to a flat area along the bank where the sample can be emptied or picked into an enamel pan with water, or into a lar with 70 percent ethanol solution for preserving the organisms. (If you do not want to preserve the organisms, and if time allows, you may be able to identify and release them at the site.)

g. If you are also collecting qualitative samples (i.e., as many organisms as you can find), be sure to collect in different habitats, such as beneath overhanging brush, in areas where the current is slower, and beneath undercut banks. You may want to label each of these samples separately so that you can match collected organisms with their habitats. If the sampling area is muddy or silty, you may need to scoop small amounts of substrate into the net and rinse water through the net a few times to remove excess mud.

I Figure 6.23. Use of a Surber sampler.

Artificial Substrate Sampling Method

a. The artificial substrate should be securely fastened to the bank, a tree, or some other permanent structure.

b. The location of the sampler should be marked so that it can be easily relocated, but does not draw the attention of vandals or other curious people.

c. Extra samplers should be placed in the stream, assuming that some will be disturbed over the course of 4 to 6 weeks.

d. During collection, care should be taken to disturb the sampler as little as possible.

e. If possible, approach the sampler from downstream and place a bucket or net under it to catch organisms that may be dislodged.

f. Place the rocks or tiles into pans partially filled with water.

g. Remove the organisms from the substrate by brushing the rocks or tiles with a soft bristle brush. Place the organisms in a 70 percent ethanol solution for later identification, or put them in a white enamel pan.

Survey Design and Methods

Survey Design

In general, sampling sites for a survey of benthic macroinvertebrates should be dispersed along the length of a river, so that changes in water quality and the resulting shifts in the aquatic community from headwaters to mouth can be noted. Comparisons could be made, for instance, among benthic communities in streams of different orders.

When the object of sampling is to evaluate the impact of a particular source of pollution or disturbance, at least three sampling sites should be tested. One site should be located just above the source to serve as a control site. Another should be located immediately below the source, and the third should be further downstream to serve as a recovery station.

(Actually, this is the survey design you should follow whether you are monitoring benthic invertebrates, or doing physical-chemical monitoring, or both.)

If sampling can only be done once or twice a year it is preferable to do it in March or April, and in late October. Ideally, samples should be collected more often (at least once per season in temperate regions).

Survey Methods

The following methods can be used to survey benthic macroinvertebrates communities. When used in conjunction with physical and chemical tests, a benthic macroinvertebrate survey can provide a good picture of the health of the river.

The methods described here generally follow a qualitative or semiquantitative approach. They have been used successfully by non-biologists to monitor water quality and identify trouble spots.

Note:

If time is limited, it is best to do the SCI and Taxa Richness procedures in order to calculate the Diversity Index.

The Sequential Comparison Index

Background Information:

The Sequential Comparison Index (SCI) is a measure of the distribution of individuals among groups of organisms; this index relates to the diversity and relative abundance of organisms. This measure is easily used by people unfamiliar with benthic identification. The SCI is based on the theory of runs. A new run begins each time an organism picked from a sample looks different than the one picked just before it.

$$SCI = \frac{\text{\# of runs}}{\text{total \# of organisms picked}}$$

Procedures:

1. Select sampling sites: The sites should be representative of the stream reach. Sampling should be done in a riffle area with a rubble or gravel bottom. Avoid areas below bridges, flow obstructions and artificial areas unless you are specifically testing for differences between these and other areas. If a good riffle area is not available, an artificial substrate may be placed in an area with visible flow.

2. Sample using a D-Frame net or kick screen. To use a D-frame net, hold the opening of the net into the current and shuffle your feet upstream from the net. Benthic macroinvertebrates should be dislodged by your feet moving on the bottom and carried by the current into the net. A kick screen requires 2-3 people, one person holding each pole and a third person kicking the substrate upstream. Three samples, or at least 300 organisms, should be collected at each station. Spend the same length of time at each station.

3. Place the samples in 70 percent alcohol preservative for later sorting. Be sure to pick clinging organisms off the net (or at least a representative selection of them if you can't get them all). Keep the samples belonging to each station separate.

4. Following the instructions below, pick organisms from the sample to calculate the Sequential Comparison Index.

   a. Make a grid of 5-7 cm squares on the bottom of a white tray. (The grid may be laid-out with a permanent marker or wax pencil.) Number the squares in order.

b. Rinse the sample of preservative, place it in the tray, and cover it with 0.5 cm–1 cm of water. Spread the sample evenly over the tray.

c. Randomly select a starting grid from which to start picking the sample. Begin picking out organisms in a random sequence. Pick all specimens from one square before moving to the next. Continue picking until all, or 50 specimens are picked.

d. Place organisms in a dish to compare each organism with the previously picked organism and record them on a work sheet using the symbols x and o (see example below). Record an "x" for the first organism picked. If the second organism picked is similar, record another "x." In the example below, the third organism picked is dissimilar to the previous organism, and so that is recorded as an "o," indicating a new run.

Example:

x	x	0	x	0	0	0	x	x	0
	1	2	3		4			5	6

Total # runs = 6   Total individuals = 10

Resulting SCI = $\dfrac{\text{# runs}}{\text{# individuals}}$ = $\dfrac{6}{10}$ = 0.6

---

e. After comparing specimens, place each in a petri dish containing similar organisms. This provides a rough sorting of the organisms into major groups to aid in identification.

f. To calculate the SCI, count the number of runs and divide by the total number of organisms.

g. Calculate an SCI for each sample. Average the samples to calculate a mean SCI for the site.

5. Determine the quantitative rating of the SCI, using the scale below. Circle the number on the data sheet that describes what you observed.

SCI VALUE		SCORE _____
4 (excellent)	0.9–1.0	
3 (good)	0.6–0.89	
2 (fair)	0.3–0.59	
1 (poor)	0.0–0.29	

Pollution Tolerance Index (PTI)

Background Information:

The Pollution Tolerance Index comes from "Save Our Streams" (The Izaak Walton League of America) and the "Citizen Stream Quality Monitoring Program" of the Ohio Department of Natural Resources. It is based on the concept of indicator organisms and tolerance levels. Indicator organisms are those organisms that are sensitive to water quality changes, and respond in predictable ways to changes in their environment. By their presence or absence they indicate something about water quality. The procedures are designed so that they can be done quickly and easily. This provides a rapid means of sampling riffle and other shallow areas in order to detect moderate to severe stream quality degradation.

The collected organisms are identified by comparing them with illustrations (Figures 6.1–6.14), or by using a key. Organisms are then classified into four groups based upon their pollution tolerance. Each of the four groups is given an index value, with the least tolerant group having the highest value. The Pollution Tolerance Index is determined by multiplying the number of kinds of organisms in each group by its index value; these numbers are then added together to form the Index. The general abundance of each kind of organism is also noted, although it is not figured into the Index.

Procedures:

1. Choose a 3 meter by 3 meter area representative of the riffle or shallow area being sampled. Use the kick seine method to sample this area. (If a seine is not available, several samples could be taken with a D-Frame net or kick screens.)

2. Three samples should be taken at each site to be sure a representative sample is collected. (Samples may also be taken from some of the other habitats at the site, such as on rocks in slow moving water, or near banks, since different organisms may be found there.)

3. Place samples in containers with a 70 percent alcohol solution for later identification. Be sure to pick clinging organisms off the net. (If you do not want to preserve the organisms, and if time allows, you may be able to identify and release them at the site.)

4. Record the presence of each type of organism collected and classify it by its tolerance (see table below). Estimate the number of each organism type (1–9, 10–49, 50–99, 100 or more) collected and record the appropriate scale on the evaluation sheet (see Data Sheet 8, page 245).

5. Calculate the Pollution Tolerance Index: multiply the number of types of organisms in each tolerance level (see Figure 6.24) by the index value for that level (4, 3, 2, or 1), and add the resulting four numbers. The following example demonstrates how to calculate an index for a hypothetical sample.

Group 1	Group 2	Group 3	Group 4
Gill Snail	Sowbug	Leech	Pouch Snail
Mayfly	Dragonfly	Flatworm	Tubifex
Riffle Beetle	Damselfly	Blackfly	Blood Midge
Caddisfly	Cranefly		
Dobsonfly			
$5 \times 4 = 20$	$4 \times 3 = 12$	$3 \times 2 = 6$	$3 \times 1 = 3$

Pollution Tolerance Index = (20 + 12 + 6 + 3) or 41

Cumulative Tolerance Index	Stream Quality Assessment
23 and above	Excellent
17–22	Good
11–16	Fair
10 or less	Poor

Pollution Tolerance Index (PTI)	Score: 41 (Excellent)

Conclusions:

Determine the quantitative rating of the PTI, using the scale below. Circle the number on the data sheet that best describes what you observed.

Pollution Tolerance Index (PTI)		SCORE _____
4 (excellent)	23 and above	
3 (good)	17–22	
2 (fair)	11–16	
1 (poor)	10 or less	

MACROINVERTEBRATE TAXA GROUPS

GROUP 1: THESE ORGANISMS ARE POLLUTION-INTOLERANT. THEIR DOMINANCE GENERALLY SIGNIFIES EXCELLENT WATER QUALITY

GILL SNAIL   STONEFLY   MAYFLY   RIFFLE BEETLE   CADDISFLY   DOBSENFLY   WATER PENNY

GROUPS 2: THESE ORGANISMS ARE MODERATELY POLLUTION INTOLERANT. THEIR DOMINANCE USUALLY SIGNIFIES GOOD WATER QUALITY.

SOWBUG   SCUD   DRAGONFLY   DAMSELFLY   CRANEFLY   CLAM

GROUP 3: THESE ORGANISMS ARE GENERALLY MODERATELY TOLERANT OF POLLUTION. THEIR DOMINANCE USUALLY SIGNIFIES POOR WATER QUALITY

LEECH   MIDGE (EXCL. BLOOD MIDGE)   FLATWORM   BLACK FLY   WATER MITE

GROUP 4: THESE ORGANISMS ARE VERY TOLERANT OF POLLUTION. THEIR DOMINANCE USUALLY SIGNIFIES BAD WATER QUALITY

POUCH SNAIL   MAGGOT   TUBIFEX   BLOOD MIDGE

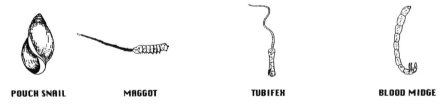

Figure 6.24. Macroinvertebrate Taxa Groups

EPT Richness

Background Information:

EPT Richness is the number of taxa (based on gross physical differences) from each of the orders Ephemeroptera, Trichoptera, and Plecoptera. This measure is known to include many species that are sensitive to water quality changes. Generally, the more EPT taxa, the better the water quality.

Procedures:

1. Choose a 3 meter by 3 meter area representative of the riffle or shallow area being sampled. Use the kick seine method to sample this area. (If a seine is not available several samples could be taken with a D-Frame net or kick screens.)

2. Three samples should be taken at each site to be sure a representative sample is collected. Samples may also be taken from some of the other habitats at the site—rocks in slow moving water, near banks—as different organisms may be found there.

3. Pick through the samples and count the number of taxa representing the three orders: Ephemeroptera, Plecoptera, and Trichoptera. The larger the number of taxa representing these three orders, the better the water quality.

4. To derive an EPT/midge ratio simply divide the total number of EPT individuals by the total number of chironomid individuals. A healthy sample should show a ratio of at least 0.75.

Conclusions:

The number of taxa, or EPT score, partially depends upon whether one sorts to order, family, genus or species. A count of 12 to 15 families indicates good water quality; a count of less than 7 or 8 is cause for concern.

It is important to use paired sites if possible when comparing sampling locations along the same river, or between rivers.

EPT RICHNESS   SCORE _____

4 (excellent)   more than 15 families
3 (good)   12–15 families
2 (fair)   8–12 families
1 (poor)   less than 8 families

Figure 6.25. Pond lilies being replaced by sedges and rushes.

Taxa Richness

Taxa richness is the number of taxa (orders, families, genera, and species) present in the sample, as identified by their gross physical differences. Taxa richness is calculated by counting the number of types of organisms in a sample. The idea behind this measure is that better water quality is generally associated with a higher number of taxa.

Diversity Index (DI)

The Diversity Index is equal to the SCI multiplied by the taxa richness. This measure is based on the theory of runs and the number of taxa found at a site. It combines into one number taxonomic richness and evenness (or relative abundance) among taxa, as measured by the Sequential Comparison Index (SCI).

DI = Taxa Richness x SCI

<u>Water Quality Rating</u>

0–8	poor
8–12	fair
12–24	good

Percent Composition of Major Groups (mostly orders)

This measure provides information about the relative abundance of different groups of organisms within a sample.

Oligocheata (worms)
Trichoptera (caddisflies)
Ephemeroptera (mayflies)
Plecoptera (stoneflies)
Chironomidae (midges)
Crustacea (crayfish, amphipods, isopods)
Mollusca (snails, clams)

Identify the organisms in each petri dish at least to order. Count the number of organisms in each of the major groups and calculate the percent composition. Graphing the results is a good way to visualize the benthic community structure at a site.

$$Percent\ Composition\ =\ \frac{number\ of\ organisms\ in\ group}{total\ number\ of\ organisms.}$$

Checklist of Sampling Equipment for Benthic Macroinvertebrates

1. White enameled pans or steel vegetable dishes (gridded if the Sequential Comparison Index is to be done)
2. Quart mason-type jars for collecting live material to place in aquaria
3. Quart mason-type jars with 70 percent alcohol solution to preserve organisms if necessary
4. Turkey baster (used to suck up small aquatic organisms)
5. Forceps
6. Meter stick (used for depth measurements)
7. Net, kick-screen, seine, or other sampler (can be homemade)
8. Vials for collecting aquatic organisms
9. Dissecting scopes
10. Vegetable brush or soft toothbrush (to gently remove attached organisms from rocks)
11. Small paint brush
12. Buckets
13. Hip waders (not essential, but good to have).

Other Stream and Benthic Macroinvertebrate Activities

Water Temperature Patterns Activity

Find the headwaters of a stream where temperature can be monitored. Measure water temperature every two hours over a day (a full 24 hours if possible). How does the temperature change throughout the day? Conduct the same pattern of temperature testing at a river's midreach. How does the pattern differ?

Measuring Velocity

1. Use a tape measure along the stream or riverbank to mark a section of the watercourse at least 20 meters in length.

2. Position someone upstream and someone downstream.

3. Release an orange into the main current at the beginning of your marked length. (An orange works well because it floats more or less in the zone of maximum velocity just below the surface.)

4. Have someone time (in seconds) the passage of the orange from the beginning to the end of the marked length.

5. The downstream person should yell when the orange floats by the end point.

6. Repeat this test at least three times and average the results. Your velocity should be in meters/second.

Calculating Discharge

Remember, discharge is a measure of the volume of water passing a certain point over a specific period of time. The method for determining discharge shown here is called the *Embody Float Method.* The formula is as shown.

$$D = \frac{WZAL}{T}$$

D = discharge
W = average width of stream
Z = average depth
L = length of stream measured
T = time for float to travel length T
A = a constant:
      (0.9 for sandy/muddy bottoms)
      (0.8 for gravel/rock bottoms)

Example:

Let's say that the measured velocity from the activity before was 0.35 meters/second.

Velocity = 0.35 meters/second

Velocity x A = mean velocity

0.35 meters/second (m/sec) x 0.8 (gravel bottomed stream in this case) = 0.280 m/sec.

Now we must measure the cross-sectional area at both ends. (The area at one end of your marked length is illustrated in Figure 6.25.)

Width of stream (meters)	Depth measurements (meters)	Area (width x depth) (sq. meters)
A–B = .5	B = .22	.11
B–C = .5	C = .34	.17
C–D = .5	D = .56	.28
D–E = .5	E = .64	.32
E–F = .5	F = .62	.31
F–G = .5	G = .60	.30
G–H = .5	H = .60	.30
H–I = .5	I = .60	.30
I–J = .5	J = .58	.29
J–K = .5	K = .36	.18
K–L = .5	L = .00	.00

Total area = 2.56 m^2

Average cross-sectional area (csa) = (beginning csa) + (lower or end csa)

$$= (2.56 \text{ m}^2 + 3.10 \text{ m}^2) \div 2 = 2.83 \text{ m}^2$$

Discharge = average cross-sectional area x mean velocity

$$= 2.83 \text{ m}^2 \times 0.28 \text{ m/sec} = 0.792 \text{ m}^3/\text{sec}$$

(Adapted from Caduto 1990)

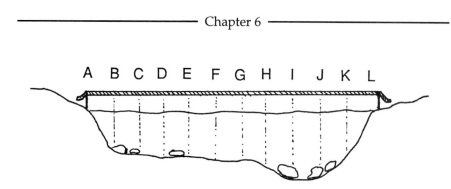

Figure 6.26. Cross-sectional area of a stream marked in .5 meter intervals. (Drawing by Mark Mitchell.)

Figure 6.27. Oil pollution coating the feathers of a pair of mallards.

Shredder Experiment

Collect leaves from one or more types of common trees, shrubs, or grasses growing along the stream. (The best time to collect is in the fall, just prior to leaf fall.) Air dry leaves for at least a week. Form leaves into 5–10 gram packs by softening or raking the leaves and putting them together in a large mesh bag.

One or two packs can be lashed to a brick with heavy string and placed in the stream facing into the current. Remove the brick after 3–4 weeks to estimate percentage of leaf remaining and counts of shredder organisms. You may want to compare results using packs of different types of leaves, such as oak, alder, maple, etc. (Adapted from Cummins, 1977.)

Leaf Pack Experiments Kit

This stream ecology kit contains everything needed to collect and study freshwater macroinvertebrates. Students use local leaves to create artificial leaf packs that attract aquatic macroinvertebrates. In this kit one finds: 6 mesh bags with waterproof labels and markers, stainless steel strainer, 14 sorting trays, 60 plastic petri dishes, waterproof invertebrate identification sheets, thermometer, scale for weighing leaf packs, 2 rulers, 12 brushes and spoons for handling organisms, 18 full-color flashcards, 6 hand lenses, Discovery Scope magnifier, nylon twine, zipper-top bags, leaf identification guide, and instruction manual with data sheets. Developed by LaMotte Company and the Stroud Water Research Center. For ordering information see page 221.

Land Use Practices and Education Activities

Take a good look at your watershed. The next time it rains, watch how the water flows over the land. Observe what the rainwater picks up on its way to local streams, rivers, and lakes. What do you suppose happens when this water enters a stream?

Our land use practices have a strong influence on the quality and quantity of water throughout the watershed in which we live. Whether we remove vegetation along the banks of a stream, or leak toxic chemicals into a river from a factory, our actions have an effect on surface waters. Any change that influences the biotic (living) or abiotic (non-living) composition of a waterway is a pollution source. The level of pollution determines what can live in the water. It also limits what we can use the water for—whether we can drink or swim in a river, or use it only to flush our wastes away.

Pollution Sources

Pollution sources are divided into two groups, depending on how the pollution enters a body of water. *Point source* pollution is waste that comes from a specific outflow pipe. Factories and wastewater treatment plants may have discharge pipes that lead directly to a waterway. These are considered point sources because they are easily identified as coming from one site.

Nonpoint source pollution does not come from a specific location. Instead, it results from the runoff of water (rainfall, snowmelt, etc.) over land. As this water passes over the ground, it picks up pollutants and carries them into local streams and rivers. Nonpoint source pollution can also result from airborne pollutants that are deposited in waterways.

Nonpoint sources can be either *rural* or *urban*. Nonpoint source pollution in rural areas usually results from such things as poor agricultural or forestry practices. Urban nonpoint source pollution is caused by the runoff from city and suburban areas.

Figure 7.1. Nonpoint source pollution comes from a variety of land use practices. (Photo from U.S. Dept. of Agriculture, Soil Conservation Service.)

While pollution from both point and nonpoint sources can be quite harmful, our ability to control the two types of sources is very different. Pollution from point sources is much easier to limit because its origin can be readily identified. If the pollution is greater than it should be, then the responsible company or plant can be contacted and asked to reduce its discharge. Stronger laws can also be passed to limit the discharge from point sources.

Nonpoint sources are more difficult to control. They are harder to identify, and sometimes nonpoint source pollution may be the result of land use practices across an entire watershed. One example is the contamination of streams and rivers by fertilizers used on lawns in suburban areas. Another example is the poisoning of rivers by agricultural runoff containing pesticides. Both types of pollution are challenging to control because there may be hundreds or thousands of widely scattered sources in a watershed. Also, nonpoint sources are not regulated by law as stringently as point sources.

A key element in controlling nonpoint source pollution is for individuals to take responsibility for their own actions. For example, people need to switch to less harmful lawn care products, and to use dangerous chemicals according to directions.

Types of Pollution

There are four major categories of pollutants, and each has different effects—direct or indirect—on the environment.

Organic Pollution

Organic pollutants come from the decomposition of living organisms, either plants or animals, and their by-products. Grass clippings, leaves, human sewage, and pet wastes are examples of organic pollution.

As discussed in Chapter 3, microorganisms use oxygen to break down organic material. When there is too much organic material in a stream, biochemical oxygen demand rises and oxygen levels fall. These conditions favor organisms that are tolerant of low oxygen levels, such as midges and sewage worms, and those that are capable of breathing surface air, such as some aquatic beetles. Mayfly nymphs, stonefly nymphs, and other life forms that require higher oxygen levels cannot survive.

The breakdown of organic material also releases nutrients into the water, particularly nitrogen and phosphorus. Excess nutrients act as fertilizer, causing an increase in the growth of algae and other aquatic plants, and leading to cultural eutrophication. The resulting decrease in light and oxygen can have serious effects on aquatic life.

Inorganic Pollution

Inorganic pollution consists of suspended and dissolved solids. These are comprised of silt, salts, and other minerals carried into streams from streets or exposed soil. Common sources of inorganic pollution include the runoff from roads that have been salted, construction sites, and plowed croplands.

Inorganic pollutants cause an increase in turbidity, leading to warmer water temperatures. (See Turbidity Section in Chapter 3.) Suspended solids also clog the gills of some fish and benthic organisms, such as mayfly nymphs. As larger particles of silt and other materials settle, they may smother fish eggs and other forms of bottom life. They may also modify the stream bottom habitat enough to make it unsuitable for many aquatic organisms, including stonefly nymphs, mayfly nymphs, and caddisfly larvae.

Toxic Pollution

Toxic pollutants are heavy metals (such as cadmium, mercury, chromium, iron and lead) and chemical compounds (PCBs, DDT, etc.) that are lethal to organisms or interfere with their normal biological processes at certain concentrations. Toxic pollutants are often produced as by-products of industrial processes. Another source are household products, such as bleach, drain cleaners, and pesticides. Street runoff and airborne contaminants can carry toxins into waterways. Standard farming practices contribute herbicides and insecticides to surface waters.

Some toxins bind to soil particles and are easily washed into aquatic habitats where they can cause an overall decline in the numbers and kinds of organisms found there. Some types of midges and aquatic worms, however, are more tolerant of toxins.

Toxic pollutants enter the food chain through organisms that process sediment, such as midges and worms. As these organisms are eaten by other animals, which are in turn eaten by larger animals, toxins accumulate in organisms further up the food chain in a process called *bioaccumulation*. In larger fish, toxins can cause lesions and deformities.

Toxic pollutants are difficult to monitor directly without sophisticated equipment. However, studies of benthic macroinvertebrate communities can provide a good indication of their presence.

Thermal Pollution

Thermal pollution is "waste heat." It often results as a consequence of using water to cool industrial or power generation processes, and then returning it at a much higher temperature into local waterways. Smaller streams are more vulnerable to thermal pollution than large rivers.

Warm water by itself is not necessarily a pollutant. It is only a problem when it creates a significant temperature difference that affects aquatic life.

As explained in Chapter 3, a rise in temperature decreases the ability of water to hold oxygen, making it more difficult for some organisms to survive. An increase in temperature also affects the metabolism of all cold-blooded life in the water—fish, insects and amphibians—possibly limiting their ability to survive under warmer conditions. Or it may create an artificial environment that some organisms come to depend upon for their survival. It may also cause insects to complete their metamorphosis earlier, which in turn affects animals, such as birds, that depend upon emergent adult insects as a food source during certain times of the year.

Let's look at the sources of pollution as we try to understand the reasons for poor water quality.

Land Use Development

How land is managed and developed influences local water quality in many ways. What we call nonpoint source pollution originates in urban, agricultural or managed forest lands as a result of a variety of causes.

Urbanization

The urban environment is a surrounding that most of us are familiar with and call our home. If we compare this area with natural areas that are protected, there are some noticeable differences.

Urban areas are places where humans live together tn relatively dense conditions and have created many changes tn the natural environment. Most of the natural vegetation and rain-absorbing soils, for example, are replaced by impermeable surfaces such as roads, parking lots, sidewalks and rooftops. Streams are often encased in concrete pipes and transformed into underground storm sewers. Natural watercourses are altered as flood-plain areas are converted to industrial, commercial, and residential uses.

As urban areas grow, the amount of pollutants washed into storm sewers and carried into local streams also grows. In separate sewer systems, untreated runoff travels through storm sewers directly into waterways.

In combined sewer systems (CSO's), sewage and storm runoff are carried by the same pipes and treated together at wastewater treatment plants. Heavy rains, however, can cause large volumes of untreated sewage mixed with street runoff to be discharged directly into waterways, resulting in severe water quality problems. This occurs in CSO's that have been designed to prevent flooding of wastewater treatment plants. To remedy the situation, some treatment plants are building retention tanks or basins to store untreated wastewater during heavy flows, to be gradually released for treatment later.

Urbanization affects water quality in many ways. As natural areas become covered by impermeable surfaces, less rainfall is absorbed by the ground; runoff increases and can create flood conditions. Storm runoff is often warmed as it passes over pavement and other surfaces, leading to warmer water temperatures in rivers and streams. Removing trees from streambanks also causes water temperatures to rise, as does the increased turbidity caused by urban runoff.

Depending on the current, the silt carried by urban runoff can scour the stream bottom in some places and smother it in other places, carrying heavy metals and other pollutants with it. Other toxins, such as used paints, solvents, cleaners, and oils, are dumped into storm drains or poured down sinks and eventually flow into local waterways. All of these pollutants have severe effects on aquatic life.

Malfunctioning septic systems along rivers, illegal sanitary connections, improperly treated wastes, and the influx of raw sewage from CSO's: all of these sources contribute bacteria and nutrients into rivers, leading to health problems downstream and cultural eutrophication.

Unsound Agricultural Practices

Crop cultivation and livestock raising are common land uses in some watersheds. Our society depends on the farmers who feed us all. However, poor agricultural practices can harm water quality as well as threaten the fertility of soils.

A variety of agricultural practices can cause rural nonpoint pollution: leaving topsoil exposed where it can be eroded; overgrazing pastures; compacting soils; leaving streambanks unprotected; overusing insecticides, herbicides, and fertilizers; allowing liquid wastes to escape from feedlots; and removing windbreaks. Many of these practices allow fertile topsoil to be carried into local streams, adding silt, toxic pollutants, and nutrients to the water, and disrupting the ecology of aquatic communities.

Salinization

Salinization—the buildup of salts in soil and water—is another serious problem, especially in irrigated areas. All water contains salts, unless it is distilled or deionized. In arid regions, natural salts concentrate in the top surface of the soil as irrigation water evaporates or is taken up by plants. After a period of time, salt levels in the soil become too high to support plant growth, and the land becomes unusable for farming. Not surprisingly, the runoff from these lands can be very saline.

Figure 7.2. Rural nonpoint pollution can be caused by overgrazed pasture or exposed topsoil. (Photo courtesy of Washtenaw Soil Conservation District.)

Excessive soil salinity can also be caused by irrigation water seeping down through the soil and raising the height of the water table. The salts in underground rocks are dissolved and carried upward as the water table rises, eventually reaching root systems and killing plants intolerant of saline conditions. This leads to wind and water erosion, since dead plants are no longer capable of binding the soil. Timber clearing in the upper watershed can also cause dramatic rises in the water table, with the same results.

Dry land salinity may be caused by reduced transpiration from an introduced vegetation type. Deep rooted trees have been replaced in the Murray-Darling river system in Australia by seasonal crops or grasses that do not pump the water into the atmosphere as efficiently.

Many farmers are now using alternatives to these destructive farming practices. Windbreaks and no-till farming helps prevent excess runoff and soil erosion. Leaving forested buffers between fields and streams helps reduce runoff dramatically (see Figure 7.3). Many farmers are converting to organic methods by reducing their use of insecticides, herbicides and inorganic fertilizers, and replacing them with natural fertilizers, such as manure, and biological methods of pest control. These alternatives are finding more appeal with farmers as they learn the benefits of protecting soils and streams.

Figure 7.3. By leaving forested buffers between planted fields, runoff is cut back dramatically. (Photo courtesy of Washtenaw Soil Conservation District.)

Unwise Forestry Practices

Depending upon the methods used, the harvesting of trees can influence water quality. Road construction, clearcutting steep slopes, and dragging the trunks of the trees into rivers to float to sawmills have all had serious effects on rivers.

One of the worst consequences of poor forestry practices is the erosion that follows the removal of trees and the ground disturbance associated with logging. This has destroyed benthic communities and ruined the spawning beds of certain fish species. At times, logging has been the culprit behind severe flooding after heavy rains, although logging debris often has the opposite effect of blocking small streams. And, as mentioned previously, the removal of trees along streams causes water temperature to rise significantly, with detrimental effects on some aquatic organisms.

Fortunately, forestry practices today are more conscious of environmental effects, and guidelines have been established to protect stream ecosystems from logging damage.

Waterway Alterations

The modification of natural water courses can affect stream ecology in many ways. Dams and other water developments provide benefits to humans, but they dramatically alter aquatic habitats.

Dams and Impoundments

Dams are built to serve community needs. Many were constructed to provide cheap hydroelectric power and control flooding downstream. Dams trap water used for irrigation, drinking water, and recreation.

Because there are few remaining free-flowing rivers, it is important to understand the ecological impacts caused by dams and their impoundments. A dam transforms a flowing stream or river into a lake-like environment. The current slows and the water warms, leading to reduced oxygen levels.

An important feature of dams is that they trap silt and sand carried by rivers. Not only does this shorten the life of the dam by filling in their reservoirs, but it changes the substrate of the river. Rooted aquatic plants flourish along the shoreline, and species of dragonfly and damselfly nymphs become more common, along with aquatic hemipterans, such as water striders and giant water bugs, and chironomid larvae. Stonefly nymphs, mayfly nymphs, and caddisfly larvae become less common.

Downstream, erosion is accelerated as water released from the dam—now effectively filtered of silt and sand—scours the riverbed. By trapping nutrients, dams also limit the diversity of aquatic life downstream.

Channelization

Another method of controlling river flow is channelization, a process in which rivers are dug out, straightened, and sometimes lined with concrete. Trees and shrubs along the river are often removed. Rivers are channelized for flood protection, farmland drainage, and sometimes for navigation.

The ecological impacts of channelization can be significant. Straightening of the river causes the speed of the current to increase, leading to greater flooding downstream. The erosion of the straightened river banks increases turbidity. In low water, the channel becomes choked with mud, an unstable substrate for benthic life. Water temperatures rise, food sources are eliminated, and the river becomes a harsh environment. Aquatic diversity drops severely.

Channelized rivers in urban areas often become a conduit for waste, and are considered a community health hazard with access restricted by tall chain-link fences. Natural, tree-lined, winding streams are converted into straight, exposed ditches—unattractive and hostile environments for people and aquatic life alike.

Wetland Drainage

Wetlands are habitats, such as marshes, swamps, and bogs, that are perpetually wet yet shallow enough to permit standing vegetation. Wetlands are some of the least understood and most threatened aquatic habitats. They have been widely drained and destroyed in the process of converting land to other uses. River channelization, the expansion of agriculture, and urbanization have all contributed to the loss of wetlands.

Wetlands are valuable resources for many reasons. They provide important habitat for waterfowl, hatchling fish, and other wildlife. They help control floods by slowing and retaining runoff. And they function as groundwater recharge sites by allowing water to percolate through the moist soil. Wetland vegetation traps the sediment in runoff and absorbs nutrients. Because of this ability to absorb nutrients, artificial wetlands have been constructed by some communities as an alternative form of sewage treatment.

Because wetlands are so important for wildlife, their loss has caused some species to decline to the point of extinction. Fortunately, recognition of the value of wetlands is growing, and many are now protected legally. In addition, many previously drained wetlands are being restored in efforts to recover the benefits of these invaluable ecosystems.

Use of Aerial and Satellite Images to Analyze Land-Use Practices Within a Watershed

Introduction

Remote sensing is the measurement of a specific property or parameter of an object or event without being in contact with that object or event, hence the word remote. In the past decade or so, remote sensing's potential as a powerful and useful tool has been realized in various disciplines.

One area where the use of remotely sensed data is increasing, and is being considered a vital key to future monitoring programs, is in environmental studies. Remote sensing can be used to study the impacts of pollution on the atmosphere, land, and water. Through remote sensing, information can also be obtained on droughts, floods, deforestation, agricultural practices, and land-river buffers.

The two major forms of remotely sensed data are *aerial photography* and *satellite imagery*. Both have benefits and drawbacks, and the system that is best to use depends on the information needed.

Despite which system is used, there are some universal strengths and weaknesses associated with remote sensing. Remote sensing allows for large scale monitoring of the environment. A single image can cover a small watershed, allowing the user to see the "big picture" and to monitor a large area instead of small individual areas. Of course, the larger the area imaged the less the resolution will be.

Another major advantage is that some forms of remote sensing allow the users to see what they can't normally see. Remote sensing can detect over a much wider range of the electromagnetic spectrum than the human eye can see. With the right system, we can detect information about objects in the ultraviolet, visible, and infrared wavelengths. For example, if we wanted to study the heat discharged from a heating plant so that we could put a stop to this form of pollution, the thermal band would detect the discharge.

One disadvantage is that with images that are of small scale, with large coverage, it is difficult to make accurate measurements. For this reason, no large coverage area should rely on remote sensing alone. Remotely sensed images should always be "ground truthed"; that is, followed up with field observations and measurements to verify the data and interpretation.

Remote sensing works on principles similar to photography. As a matter of fact, conventional photography is a form of remote sensing. The remote sensing system has a sensor that is able to detect various wavelengths. All sensors have one, or many, detectors which respond to energy striking them. The intensity of the energy flux at the detector is recorded as an index of the "brightness" of the object in that spectral band. The spectral bands may include the visible bands, which all people with sight "see," but also many other bands. The sensor detects through the exposure

Figure 7.4. Remotely sensed information on diverse land use practices around Lake St. Clair.

of a detection surface that is sensitive to specific wavelengths, as with conventional films sensitivity to the visible bands.

After this, all that is left to do is to form the image. The two main forms of remotely sensed data, as stated earlier, are aerial photography and satellite images. Because each satellite is launched once, and used for many years, it is important to design the system so that it records over a range of spectral bands that will meet the needs of many users. For this reason satellites today, with the Landsat series and the French Spot being the most common systems, are able to detect in many different bands.

Water quality can be measured using the National Sanitation Foundation's nine water quality tests: dissolved oxygen, fecal coliform, pH, biological oxygen demand, temperature, nitrates, phosphates, turbidity, and total solids. Although remote sensing cannot directly measure these parameters, except turbidity and temperature, it can detect other river characteristics that would indicate something about the levels of one or more of these parameters. It can also be used to determine the land use of the surrounding terrain, which can then be used to infer the measurements of certain parameters.

There are also *indirect* methods for detecting some of the other parameters, such as nitrates and phosphates, by detecting changes in phytoplankton. Also, through the relationship turbidity and temperature have with the other parameters, knowledge of these two gives us insight into the others. Of course, all inferred measurements should be verified through ground truth studies.

Now that you have had a brief introduction to how remote sensing works and its applications to water quality management, you are ready to apply these principles.

Activity

7.1

Your River From Beginning to End

Concept

This activity is designed to provide students with a spatial knowledge of their river and watershed as a foundation for understanding the processes and factors that determine water quality. Teachers may lecture about critical factors, such as runoff, and maps may provide a visual representation of important concepts, such as locations on a river, but neither of these techniques compares to the benefits of actually seeing the whole river system.

In most cases it is impossible for a class to visit the whole river, from headwaters to mouth, and even if it were possible to visit a watershed, it does not always offer the best perspective. Students need to gain an understanding of the river as a whole; they need to see the river's flow and its tributaries; they need a spatial view of the river. Aerial photographs and satellite imagery offer this viewpoint. By looking at images of the whole river, students can gain a better understanding of watersheds, headwaters, tributaries, stream order, drainage, discharge, and many other factors and characteristics.

Supplies

You will need aerial photographs or satellite images of your watershed. Sources of aerial photography can be obtained by requesting the pamphlet, "How to Obtain Aerial Photographs," available from U.S. Geological Survey, User Services Section, EROS Data Center, Sioux Falls, SD 57198; Telephone (605) 594-6151. Landsat images are available for the 50 states and for most of the earth's land surface outside the U.S. A useful brochure, "Landsat Products and Services," is also available at the same address.

Your local watershed council, planning bodies, universities or state water agencies might have aerial photographs or satellite images available for classroom use. Aerial photographs and satellite images could be photographed on slides (with permission), and reproduced in color into 11" x 17" (28 x 43 cm) prints on some commercial copying machines, for class use.

For this and the following activities, you will also need 8.5" x 11" (21.5 x 28 cm) transparencies and photographic tape. (Both items are available at office supply stores. You can use ordinary adhesive tape, but photographic tape is desirable to avoid tearing your images.) You will also need a transparent 1/4" (4 mm) grid. You can make this by drawing the grid on ordinary paper, then using a copy machine to transfer it to one of your transparencies.

Procedure

A river begins at its headwater and flows downstream to its mouth. Headwaters are the river's source and are located the farthest upstream. The mouth is where the river ends. Most rivers will end when they flow into another water body, such as a lake, ocean or another river. Locate the headwater and mouth of your river on the aerial satellite image of your watershed. Now look at the river channel; is it long, windy, are there any distinctive features?

I **Figure 7.5.** Students examining their local river from beginning to end.

One important factor that will affect river water is the river's source. The source of a river is usually either underground water coming to the surface as springs, rainfall or snowmelt, wetland drainage, glacier meltwater, or the outlet of a pond or lake. By looking at the headwaters can you determine the source of your river?

Looking at the characteristics of the surrounding land should help you with your determination. For example, if the source is wetland drainage there should definitely be a wetland in the proximity of the river. You can verify your final decision with county records or some other source of information about your local river. This information can be obtained from the watershed council, a planning agency, or other appropriate authority.

Another factor that influences a river's characteristics is the land use found along its banks, such as the presence or absence of vegetation. What is the land use on the banks near the headwaters; near the mouth? How does the land use change as you move down the river? How would you expect these different uses to affect the river?

When studying rivers and the organisms that live within them, it is important to consider water flow. Although we cannot measure rates of flow from images, sometimes we can determine flow to the extent of whether a river is calm or turbulent. Aerial photographs would be best for this activity but satellite images can be used also. By looking at the image you might be able to detect rougher water. It may look different if it is very rough, e.g. "white water". Calm waters tend to look flat and uniform, while very rough waters will look bumpy and have a definite heterogeneous appearance.

If your image does not offer great enough resolution to determine this, you may still get a sense of the flow by looking at the river as a whole. Can you see the bottom of the river? Are there objects, such as rocks, protruding from the water that would result in turbulence? Does the width of the river vary considerably? If the river is wide and then narrows, the same volume of water will have to get through a smaller area. This would result in a more rapid, turbulent flow.

By now you might have noticed that there are other water inputs, such as streams, into the river you have been studying. All of these inputs and the river make up a branching network known as the river system. This system extends from the headwaters to the mouth.

To describe the position of a stream in this river system a method called *stream order* has been devised. A single stream, or river, that has no inputs is called a *first order stream.* When two first order streams merge the resulting stream, which usually continues in the path of one of the original steams, is called a *second order stream.* Therefore, a single stream is a first order stream from its origin to the point where it joins with another first order stream.

When two second order streams come together the resulting stream is *third order*. Third order can only result from the merging of two second order streams; if a second order and a first order stream meet, the resulting stream is still second order. The joining of two third order streams results is a fourth order and so on. The Mississippi River, around New Orleans, is a twelfth order river.

Using your images, determine the order of your river, at the point of your monitoring station. Since you cannot write on these images, you might want to overlay a transparency and trace the river system. Then you can do the numbering on this sheet.

Activity

This Land is Your Land, This Land is My Land

Concept

Pollutants, or that which is harmful to rivers, can be introduced to a river system through point sources, such as pipes, or non-point sources, such as runoff, or atmospheric deposition, in the form of either rain, snow, or dry deposition. Although some pollutants occur naturally, such as those from volcanic eruptions, most are the result of human activities. Very few people consider that the wastewater from washing their car might end up in a river and kill fish; that a parking lot replacing an abandoned field means more runoff containing harmful debris, such as sediment and toxins; or that a farm's fertilizers might impact a life in a river through the depletion of oxygen.

This activity is designed to arm students with the knowledge they will need to make a difference to improve their environment. By studying different land uses in their watershed, students should develop an understanding of how these uses affect water quality and the role that they, their families, their school, their neighbors, and others play in land use decisions. The primary goal of this activity is *not* just to make students aware of human impacts, but to make them aware that they are not powerless, and that they can help improve the environment.

Supplies

Refer to the "Supplies" section in Activity One.

Figure 7.6. Students locating point and non-point sources of pollution on their river.

Procedure

Working either in pairs or larger groups (depending on the availability of images), obtain an image that contains your watershed, a transparency, photographic tape, and a pen from your teacher. Place the transparency over the part of the image that you are interested in studying. Once you have properly placed the transparency, tape it down. If you decide that you need to move the transparency, carefully remove the tape starting with the end on the transparency and not on the picture—this should prevent you from tearing the picture. Once you have the transparency properly placed, outline your watershed.

Land use is an important determining factor for water quality. Whether it be residential, agricultural, or industrial land, the activities that take place on land will be reflected in water quality. Different land uses will contribute to different types of water quality problems. For example, farmland adds nutrients and pesticides to rivers. Highway bridges result in the introduction of oil, gasoline, or salt to rivers. Septic tanks in residential areas cause a increased biochemical oxygen demand. Factories discharge warm water directly into the river.

Pollutants will also come from the atmosphere in such forms as acid rain. These atmospheric pollutants are due to human activity also, but since air moves freely between watersheds, it is beyond the scope of this activity to determine the origin of these pollutants. However, when looking at the different land uses in your watershed you should consider how they could add to atmospheric pollution that may end up polluting rivers miles downwind.

Since land use is so important, it is essential that you know the land uses in your watershed and the impacts they have on water quality. Starting at your river's headwaters and working to its mouth, identify the major land uses along the river. What would you expect the water quality to be along each stretch? Consider the characteristics that are related to each major land use that may or may not be present. What differences would you expect in water quality as a result of the presence/absence of these characteristics?

Activity

 7.3 | # A Changing World

Concept

Due to natural and human-caused factors, our environment is constantly changing. Some of these changes will be temporary, others will be so minor that they will go unnoticed, and others will be drastic enough to be permanent. Although these changes are not always for the worse and have varying effects on different species, an awareness of them is essential for a fuller understanding of our closely integrated environment.

This activity offers students the opportunity to see the changes that have taken place in their watershed over time. Students will have the chance to speculate about possible causes of these changes, impacts, and ways to prevent future undesirable changes. Their awareness and knowledge of their own town's history will be heightened.

Supplies

Refer to the "Supplies" section in Activity One.

Procedure

Roads are being repaired and constructed all around cities. For this road work, trees are cut down and grass is dug up, exposing soil. Wind and rain carry dirt away. Some soil ends up in the local river, causing the water to become more turbid, warmer, and cloudy, suffocating some organisms.

Figure 7.7. Through remotely sensed data students can determine the impact of changes in land usage on water quality.

Factories and industries are being built. Many are built near rivers since they require water to cool their machines. As the water flows around the machinery, it is warmed. In some cases this warm water is released directly back into the river where it can interfere with the biological processes of aquatic organisms, such as reproduction and metabolism.

Commercial facilities build parking for their employees and customers. Unlike soil, concrete is impervious to water. The result is increased runoff that may contain harmful pollutants, such as oil from cars.

By making some compromises and taking a few preventive measures, we can lessen the impact of most changes. For example, when building roads, construction workers can cover the exposed dirt with some type of mesh to limit the amount of erosion. Factories can use cooling ponds or towers for the water that they warm, allowing it to cool before returning it to the river. Parking lots can be designed so that runoff is routed to holding ponds where the water has a chance to percolate into the ground.

Working in small groups, obtain images of your watershed over past years. How long ago was your earliest image taken? Spend some time familiarizing yourself with the image. Can you find your school in any of the images, or was it built in later years? What about other common features in your watershed? Are there buildings or other features that are evident in earlier images but not found in later ones?

Compare the images consecutively, noting any new developments or expansions of older developments. How might these changes affect water quality? Also note the disappearance of any features and replacements, such as a housing development being replaced by a mall. Be sure to note the relative distance from the river of these features, as this might influence their impact.

Also, the land use between the feature and the river could have an effect, such as the contrast between impervious and pervious. Discuss that land use and how it might modify the effects of the outer land use. Discuss the possible impacts these land uses could have on the quality of the river water on both a broad scale and a specific scale, using the nine water quality tests as a guide. Does it appear that any preventive measures were taken to lessen these impacts? An example of this would be a farmer leaving a natural barrier or vegetation between crops and the river. Discuss any suggestions you have for preventive measures. Be sure to state why you believe that they are beneficial.

Take your earliest image and tape a transparency to it. (If you need to move the transparency, pull the tape off from the transparency first so that you will not tear the image.) Writing on the transparency, differentiate the impervious from the pervious land using different color pens or some other marking system. Now overlay the grid on your transparency. Find the proportions of impervious and pervious areas. Do this by counting the number of grid squares that fall within the total area of interest. Let this number equal "t". (For those squares that are not totally in the area of interest, estimate the amount to the nearest quarter square.)

Now estimate the number of squares that are impervious, again making all measurements to the nearest quarter square. Let this number equal "n". The proportion is "n/t". Repeat this for the pervious land.

Once you have determined the proportions for the first image, repeat this process for all images. If you can, without it being too confusing, do all the work on the same transparency so that you can visualize the change. Was there a substantial change? What changes in water quality would you expect to result?

Activity

Clarity through Cloudiness

Concept

Remote sensing is a powerful tool for monitoring the environment on a large, efficient scale. The techniques of remote sensing can be applied to many different situations, but due to its detection mechanisms not all problems can be researched by the measurement of conventional parameters. Remote sensing works on principles similar to photography. It works by detecting wavelengths along the electromagnetic spectrum that are emitted or reflected by objects. This detection usually entails the exposure of a film that is sensitive to specific wavelengths.

Turbidity can be detected in the visible band, amongst others. Since turbidity is also related directly and indirectly to most of the other eight water quality tests, information about these other parameters can be inferred from turbidity data. In this activity, images will be examined for their relative turbidity levels, from which students will be able to make a statement about the overall water quality. In addition, if desired, students may also "ground truth" turbidity by taking readings on the day of a satellite or airplane overpass.

Supplies

Refer to the "Supplies" section in Activity One.

Procedure

Turbid water is water that contains suspended sediment, resulting in a cloudy appearance (see Chapter 3). The sources of this suspended sediment can be both external and internal to a river. External sources are activities that cause vegetation and soil to be disturbed, such as building construction. These allow erosion to occur and runoff to carry soil into the river. Internal sources are anything that stirs up bottom sediments in a river, such as bottom-feeding fish like carp. The more turbid the water, the cloudier it will appear.

Obtain the images of the portion of your watershed that you will be studying. Overlay the transparency, securing it with photographic tape. (If you need to move the overlay, remove the tape from the overlay first to prevent tearing the image.) Carefully delineate the regions of different turbidity by tracing the edges of the areas that appear different. After you have divided the region into different levels of turbidity, number them according to increasing turbidity. Clear water should be number one.

Using your knowledge of land use and turbidity, discuss your results. What causes certain areas to be more turbid than others? Based on the relative turbidity, what can you conclude about the other eight parameters?

Now that you have determined the relative values of turbidity and closely related parameters, obtain actual water quality data from your study region. You can obtain this data by testing the river yourself, but if your study area is large or there are too many different turbidity regions this might be quite difficult. In this case, contact other schools in your watershed that are also monitoring water quality. If their testing site is within one of your turbidity regions you can use their data as ground truth. If this does not work, contact the local agency in charge of water quality monitoring and obtain any applicable data they might have. Compare it to your hypothesized results. Were you right in your relative assumptions? For those areas where you were wrong, can you tell why by looking at the image?

Activity

A Step Above

Concept

One of the government's main functions is to protect and sustain our natural resources. With these responsibilities, it should come as no surprise that the government is active in monitoring our rivers and their quality. Although the monitoring of individual rivers is done on a state and local scale, the national government is concerned about ensuring that our waterways are clean and safe.

This activity is designed to give students a better understanding of the whole regulatory process. By visiting a local or state agency, students will have the opportunity to see how their government is meeting the need to monitor our rivers, to interpret data, and to develop and implement methods to protect water quality. Students should gain an understanding and appreciation of the complex, intricate work done by their government.

Through the knowledge of how the government works, students will hopefully realize that they can be influential in their government by contacting the right people and using the proper approach to issues. This activity should leave students with the realization that there are no simple answers to most of our problems, but through the concern and help of many, solutions can be found.

Procedure

Spend about half a class period explaining remote sensing and water quality to your students. Be sure to discuss the reason why rivers are monitored and the role that the government plays in this monitoring, and why. Also discuss the principles of remote sensing.

Contact the local or state agency that is in charge of monitoring water quality. Set a time for your class to visit and make the agency aware of your goals and area of interest. Let them know that you want your students to learn about the whole process of monitoring river water quality, from the actual collection of data, to the analysis of the information, and implementation of the plans. Also indicate that your emphasis is on remote sensing and therefore your students would need to hear about whether it is used by the agency, and why or why not.

Each student should develop three questions that he/she can ask during this visit. These questions can be on anything related to how the agency regulates water quality, the difficulties of its job, the role citizens can play, and the agency's successes. One question should be on remote sensing and its application to water quality planning and management. If the agency does not use remote sensing find out why, if they ever plan to, and if any regulating in your state is done with the aid of remote sensing. After your visit, have your students write a one-page paper on what they learned. This activity might also help students reflect on potential employment opportunities, using the skills related to remote sensing and water planning.

Additional Educational Activities

Activity One: Stream Survey

Often we can learn much about the health of a river by observing its appearance and general surroundings. A sample stream survey is shown in Figure 7.8.

This activity should be done at the same time as the physical and chemical tests described in Chapter 3; that way, the results of the tests can be linked to observations made during the stream survey to identify the likely causes of water quality problems.

Activity Two: Contacting Decision-Makers

After students have collected and analyzed all of their results, they will probably identify a potential water quality problem from one or more of the tests. The next step is to determine what could be behind these results. By referring back to the description of each test in Chapter 3, they should be able to identify general causes. However, there is far more information available to students if they are motivated to find it.

Modified from Michigan Department of Natural Resources
Surface Water Quality Division

STATION NUMBER _____ INVESTIGATOR(S) _____ DATE _____

BODY OF WATER _____ LOCATION _____ TIME _____

COUNTY _____ TOWNSHIP _____ T _____ R _____ S _____

STREAM TYPE: □ Coldwater □ Warmwater REASON FOR SURVEY _____

SURVEY CONDITIONS

WEATHER: □ Sunny □ Partly Cloudy □ Rainy AIR TEMP. _____ WATER TEMP. _____

LOCAL LAND USE: □ Urban □ Suburban □ Agricultural □ Grassland □ Forest □ Other _____

SURVEY REACH LENGTH:_____ft. % STREAM SHADING: _____ CHANNELIZED: □ yes □ no

DAM UPSTREAM: □ yes _____ft. □ no

STREAMBANK VEG.: □ Barren □ Grasses □ Herbaceous □ Brush □ Deciduous □ Conifer □ Other _____

    Est. % coverage _____ _____ _____ _____ _____ _____

    Vegetation height (ft.) _____ _____ _____ _____ _____ _____

BANK STABILITY: □ Stable □ Slightly Eroded □ Moderately Eroded □ Severely Eroded

                                         BANK MATERIAL _____

                                         UNDERCUT BANKS □ Yes □ No

AVE. STREAM WIDTH _____ft. AVE. STREAM DEPTH _____ft. SURFACE VELOCITY_____ ft./sec

                                                     EST. FLOW _____cfs

BANKFULL HEIGHT_____ft. BANKFULL WIDTH _____ft. CHANNEL SLOPE _____ ft./mile

                                                                (from MDNR)

CHANNEL X-SECTION: □ Rectangular □ V-Shaped □ U-Shaped □ Other _____

                       % BANK SLOPE: □ Steep □ Moderate □ Slight

TURBIDITY: □ Clear □ Slight □ Turbid SURFACE OILS: □ None □ Some □ Lots

WATER ODORS: □ Normal □ Sewage □ Petroleum □ Chemical □ Other _____

UNDERSIDES OF IMBEDDED RUBBLE BLACK? □ Yes □ No (Indication of Oxygen available)

INORGANIC SUBSTRATE	FLOW VELOCITY	CHARACTERISTICS OR SIZE	PERCENT IN SAMPLING AREA	ORGANIC SUBSTRATE	CHARACTERISTICS OR SIZE	PERCENT IN SAMPLING AREA
INORGANIC:				ORGANIC:		
BOULDERS >3	fps	> 10 inch dia.	_____	MUCK-MUD	Black, very fine organic.	_____
RUBBLE 2	fps	2.5-10 inch dia.	_____	PULPY PEAT	Unrecognizable plant parts	_____
GRAVEL 1	fps	0.1 - 2.5 inch dia.	_____	FIBROUS PEAT	Partially decomposed plant material	_____
SAND 0.7	fps	0.002-0.079 inch dia.	_____	DETRITUS	Sticks, wood, coarse plant material	_____
SILT < 0.4	fps		_____	LOGS, LIMBS		_____
CLAY	100%	Slick Texture	_____			100%

SITE SUBSTRATE COMPOSITION: % Inorganic _____ % Organic _____

Figure 7.8. Stream survey used for the Rouge River, Michigan.

In the United States, a number of agencies have been given the responsibility of monitoring, and in some cases managing, local waterways. The Clean Water Act of 1972 set water quality standards for rivers that industries and municipalities must observe. Worldwide, many other countries have national standards for water quality, or are in the process of creating them.

If students want to find out who has more information about their river, and perhaps the power to clean it up, there are several places to which they can turn. These are the steps involved:

1. Find out which, if any, agencies are concerned with river water quality. This can be done by contacting your mayor, governor, or any other local governmental leader. If you describe your interests carefully, you should be directed to a specific office. (If there is no office, perhaps you can work to have one created!)

2. The next step depends on your goals. If you only seek further information, then you should be able to get what you want by speaking with the appropriate government official. You can then share this information with the class.

   On the other hand, if you want to undertake some problem-solving to see your issue solved, this will take more work. It will probably require research and organizing not within the scope of this book.

Activity Three: Stenciling Storm Drains

Runoff from urban areas creates many water quality problems. Some students in the Pacific Northwest region of the United States have found a creative way to educate people about these problems, particularly the improper disposal of used car oil and antifreeze. Because people sometimes pour these and other hazardous wastes down storm drains or into the street, some schools have begun stenciling their messages right on the drains. Here's how they do it:

The first step is to locate and map storm sewer drains in your area to learn how they connect to local waterways. Colored food dye can be poured down the drain to determine exactly where it enters the stream. (This can also be used as a graphic demonstration of the connections between storm drains and streams.)

Next, draw and cut out stencils that say something like "DUMP NO WASTE . . . DRAINS TO STREAM" or "DON'T DUMP . . . FISH BELOW." Then, spray paint these messages on storm drains (after first obtaining permission from local municipal authorities; it would be good to obtain permission at the same time to restencil every few years also.) Some groups

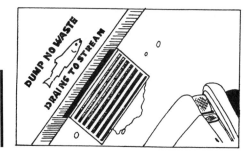

Figure 7.9. Stencil on storm sewer informing residents that waste deposited in the sewer runs directly to the river. (Adopt-A-Stream Foundation © 1988. Used with permission. All rights reserved.)

have stenciled their drains as part of a broader public education campaign that included giving out brochures.

Activity Four: Home Hazardous Waste Product Survey

This is an excellent activity to demonstrate the amount of household toxics that people unknowingly put into the environment. After these products are bought, they usually end up in landfills after being thrown away, or in rivers after being poured down the sink. In both cases, they continue to be toxic and contaminate groundwater or surface water.

Students are assigned to inventory the potentially hazardous products in their homes (in the garage, basement, under the sink, etc.), with their parents' assistance, if possible. Then, they are asked to tabulate the results with their classmates. The handouts at the end of this chapter can help with these tasks. The results will reveal the number of containers of hazardous materials found in the average household.

After estimating the number of houses in a community, and averaging the amount of hazardous containers in the entire area, it is easy to see how our homes can contribute more chemicals to the environment than a factory. In some communities, students have found that the number of hazardous waste containers in the average home is between 40 and 80.

Can Some of Your Household Products Harm You?

TOXICITY RATING	LETHAL DOSE FOR 150 lb. HUMAN	HOUSEHOLD PRODUCTS . . .
Practically Non-Toxic	More than 1 quart	foods, candies, lead pencils, eye makeup.
Slightly Toxic	1 Pint to 1 Quart	dry cell batteries, glass cleaner, deodorants and anti-perspirants, hand soap.
CAUTION: Moderately Toxic	1 Tablespoon to 1 Pint	antifreeze, automotive cleaners, household bleaches, many detergents, dry cleaners, most oven cleaners, many general cleaners, most fuels, lubricating oils, most stain and spot removers, many disinfectants, floor polish, shoe polish, most paints.
WARNING: Very Toxic	1 Teaspoon to 1 Tablespoon	toilet bowl cleaners, some deodorizers engine motor cleaners, some fertilizers, some paint brush cleaners, some paint and varnish removers, fireworks, some mildew proofing, air sanitizers, some paints, lacquer thinners, many pesticides: DDT, chlordane, heptachlor, lindane, mirex, diazion, malathion, diquatdibromide, endothall, 2, 4D.
DANGER: Extremely Toxic	1 Drop to 1 Teaspoon	some of the insecticides, fungcides, rodenticides, herbicides: aldrin, eldrin, bidrin, paraquat, some fertilizers and mercury batteries.
Super Toxic		a few pesticides like: paroxon, phosdrin, parathion, isobenzan, pyrazoyan.

Directions: Look through your house to find the products that are used in each of the following areas. Read the labels to determine if the product you have is potentially hazardous. If it is, put a check in the blank to the left of the item. Estimate the number of containers, and write that number in the blank that follows the item. Circle the products that you personally use. Please check with a parent before starting this project: They may wish to learn from the survey, too!

KITCHEN:
_____ oven cleaner _____
_____ floor cleaner and wax _____
_____ disinfectant cleaner _____
_____ ammonia _____
_____ scouring powder _____
_____ bleach _____
_____ other _____
Total Containers: _____

LIVING ROOM: (Look in a nearby cupboard or closet)
_____ rug cleaner _____
_____ furniture polish _____
_____ air freshener _____
_____ other _____

Total Containers: _____

LAUNDRY ROOM:
_____ bleach _____
_____ spot remover _____
_____ detergent _____
_____ other _____

Total Containers: _____

LAWN:
_____ weed killers _____
_____ insecticides _____
_____ fertilizers _____
_____ other _____
Total Containers: _____

BATHROOM:
_____ disinfectant tub/tile cleaner _____
_____ drain opener _____
_____ toilet bowl cleaner _____
_____ medicine _____
_____ other _____

Total Containers: _____

WORKBENCH:
_____ paint _____
_____ varnish _____
_____ glue _____
_____ paint thinner _____
_____ furniture stripper _____
_____ wood preservative _____
_____ other _____
_____ photographic supplies _____
Total Containers: _____

GARAGE:
_____ oil _____
_____ antifreeze _____
_____ rat poison _____
_____ gasoline, kerosene _____
_____ other fuel _____
_____ pool chemicals _____
_____ other _____
Total Containers: _____

OTHER PLACES AND ITEMS:
_____
_____
_____
_____
Total Containers: _____

TOTAL NUMBER OF HAZARDOUS MATERIALS IN YOUR HOME:
_____

Class Data Sheet

PART I:

Tally the number of household hazardous materials in each category from all the households in your group. Then total again to find the class's entire collection of toxic containers.

GROUP   CLASS

Kitchen _____ _____

Living Room _____ _____

Laundry _____ _____

Lawn _____ _____

Bathroom_____ _____

Workbench _____ _____

Garage _____ _____

Other _____ _____

TOTAL: _____(a)

PART II:

Number of households surveyed: _____ (b)

Average number of hazardous materials per household: _____ (a - b = c)

Number of households in your community: _____ (d)

Your estimate of the number of household hazardous materials in your community: _____ (c × d)

*Alternatives to Household Hazardous Waste

Common toxic household cleaning products can often be replaced with non-toxic alternatives:

Bleach: oxygen bleaches, sun-bleaching of clothes.

Deodorizers: an open box of baking soda, herbal arrangements, cedar chips, cinnamon and cloves.

Drain Cleaners: using a plunger, followed by 1/4 cup of baking soda and 1/2 cup of vinegar. Allow to sit for 15 minutes, then rinse with boiling water. Mechanical methods can also be used—one method employs a long flexible metal rod called a "snake". The best alternative is prevention, particularly hair removal, band by inserting into the drain a rubber screen to catch material.

Dusting: use 1/4 cup of white vinegar in one quart of water; apply with a soft cloth.

Furniture/wood polish: rub with 1 tablespoon of lemon oil mixed in one pint of mineral oil.

Glass cleaner: use 2 tablespoons of vinegar in 1 quart of water.

Mildew stain remover: use of vinegar solution composed of 1/2 cup vinegar and 1 quart of warm water.

Spot cleaning carpets: apply club soda immediately, blot dry, and repeat. Sprinkle with cornmeal or cornstarch and vacuum after 30 minutes.

Toilet cleaner: use baking soda, or vinegar, or non-chlorinated scouring powder.

*These alternatives are taken from the *Pudget Soundbook* (1991) produced by the Marine Science Center, 18743 Front St. NE. P.O. Box 2079, Poulsbo, WA 98370.

Landfill

Most solid waste, including improperly discarded hazardous materials, from households, schools, and businesses (paper, cans, glass) eventually ends up in the local licensed sanitary landfill. These large pits are usually lined with clay and/or industrial strength plastic to prevent **leaching** of contaminated water. In the past, many communities burned this waste on a daily basis, however air quality was greatly affected by this practice. Presently, the methods of compaction and burying are used. At the end of each day, the waste is smashed and covered. The process is repeated each day. Once the pits reach full capacity, they are completely covered and sealed with clay, which is designed to prevent water from percolating through the waste.

One major problem associated with landfills is that the clay and/or plastic layers do not always prevent contaminated water from leaching through. Another problem arises from the disposal of hazardous waste in local sanitary landfills. Many of these landfills were not designated to contain certain hazardous materials. Contacting your local sanitary landfill and/or hazardous waste disposal site may give additional insight.

Discussion Questions:

1. Which household hazardous wastes might be disposed in a sanitary landfill? How could these household hazardous wastes contaminate the environment?

2. What are the basic guidelines for sanitary landfills in your state and who is responsible for enforcing those regulations?

3. What are the consequences of an improperly built landfill?

LANDFILL

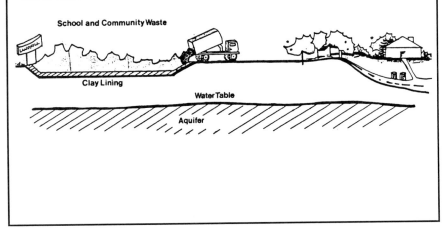

School and Community Waste

Clay Lining

Water Table

Aquifer

Septic Systems

Many communities, as well as some drive-in theatres and shopping centers, operate on private septic disposal systems. Traditionally, these systems handle many of the same type of substances that are received by city wastewater treatment (waste water from sinks, drains and pipes, toilets). These communities are usually in areas that cannot be serviced by city waste water treatment due to distance or unavailability of hook-ups. This system consists of two basic elements: a septic tank and an absorption field. A **septic tank** is simply a tank that is buried in the ground to collect and treat the sewage. Wastewater flows into the tank where it is broken down by **aerobic** (utilizing oxygen) and **anaerobic** (without oxygen) bacteria. To continue functioning properly, the tank may need to be emptied periodically, by pumping out the built up solid waste or sludge. That material is sent to the local sanitary landfill.

The solid wastes settle to the bottom and the **effluent** (wastewater) flows out to the overflow pipe, it is carried by gravity through the pipe to the absorption field. An **absorption field** consists of perforated tiles laid in gravel or crushed stone. The wastewater or effluent travels through the perforated pipes and trickles into the soil. The soil acts as a filtering system where additional aerobic decomposition takes place. In a properly constructed septic system, the effluent should be free of organic waste by the time it reaches the water table. Some household chemicals (ie. concentrated drain cleaners) may destroy the bacteria in a septic tank. Without the bacteria decomposing the waste, the treatment ability of the septic system decreases.

Discussion Questions:

1. List 5 household hazardous materials that end up in a septic system, either when used or disposed. Would the septic system be able to treat that material?

2. What are the guidelines in your community for septic system construction and maintenance? You may want to contact the city or county health department.

3. If these guidelines are not met, what may be the possible consequence?

SEPTIC SYSTEM

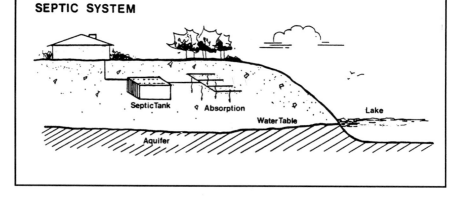

Municipal Wastewater Treatment

The most common system for treating household wastewater (from sinks, tubs, toilets, floor drains) on a municipal level is sewage or wastewater treatment. Through a series of pipes the raw sewage from homes, businesses, and industries is carried to the wastewater treatment plant. Here the solids are removed and the effluent treated, and discharged into nearby rivers or streams. Some wastes are not allowed to be placed in the system if they might damage the operation of the treatment plant.

There are two kinds of sewer systems: separate and combined. A **separate system** collects raw sewage in sanitary lines and delivers it to a wastewater treatment plant. Rainwater and materials from streets flow into separate storm water lines and are normally discharged to a nearby river or stream. If more sewage is delivered to the plant than can be handled, some raw sewage may bypass the plant or be held in a retention basin for later treatment. In a **combined system,** sanitary lines and storm water lines flow together into the wastewater treatment plant. During heavy rain storms, some of the combined rainwater and sewage may bypass the treatment plant and go directly into a stream or river without any treatment at all. In some cases it can be stored in a retention basin for later treatment.

PRIMARY STAGE

When the sewage arrives at the plant, it flows through a grit chamber which settles out large grit and particles. From the grit chamber, the wastewater flows through a series of screens which collects and shreds medium-size debris such as paper. The openings of the screens diminish in size, collecting smaller and smaller bits of material. The material collected by the screen is taken to a sanitary landfill and the wastewater which passes through the screens flows to sedimentation tanks. In the sedimentation tanks, the velocity of the water is greatly decreased allowing solid material to settle to the bottom where it is scraped into hoppers and transferred to sludge treatment tanks for further decomposition and disposal in a landfill or incinerator. The effluent is now ready for secondary treatment.

SECONDARY STAGE

In the secondary stage, up to 92 percent of the organic material remaining in the effluent is digested by aerobic bacteria utilizing oxygen. Basically, the effluent is brought into contact with large numbers of live bacteria which consume a majority of the organic matter. Sprayers or bubblers add oxygen to the water to maintain a healthy environment for the bacteria. The wastewater is then sent to a final settling tank where chlorine is added to disinfect the water before returning the treated water to the river.

TERTIARY STAGE

Some communities treat their wastewater with an additional third stage. Tertiary treatment systems are used to remove additional organic and inorganic material (nitrogen, phosphorus) from the effluent and to increase the quality of the plant's effluent. Sand filtration or spray irrigation systems filter most of the remaining small quantities of particles and waste material out, providing around 99 percent organic and inorganic waste removal. Some tertiary systems are so efficient that the effluent is safe for human consumption

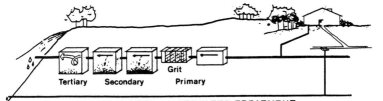

Tertiary   Secondary   Primary   Grit

MUNICIPAL WASTEWATER TREATMENT

Discussion Questions:

1. State what you feel would be the advantages and disadvantages of both the separate and combined wastewater treatment systems. Which type of system does your community have?
2. What would be the consequence of releasing raw sewage into a stream or river?
3. Are there any materials that cannot be handled by this type of system?
4. List 5 household hazardous materials that end up in a municipal wastewater treatment plant when used or disposed. How might they effect the system?

Groundwater

Groundwater is the water which accumulates below the ground's surface. It percolates, or trickles down through porous soil. In some locations there are large amounts of water, called aquifers, trapped within rock formations or found in underground sand and gravel deposits. It is from these aquifers that we receive the majority of our drinking water. A variety of sources have the potential to contaminate this sub-surface water.

Aquifers and groundwater are a renewable resource made possible by the **hydrological cycle.** Moisture evaporates from the surface at the ground and from bodies of water (lakes, rivers, and streams) or evapotranspires from trees and shrubs. When the density of the vapor reaches a critical point, the vapor begins to form clouds. If this process continues, the clouds will release moisture in the form of rain. The rain will either be absorbed by the ground and percolate to the water table and recharge the aquifer, or it will run off the surface to a lake, river or the ocean. The surface area in which the water drains to a common location is called a **watershed** or drainage basin; it can be imagined as the valley between two adjoining roofs. Watersheds may be protected by local, state and/or federal legislation. The hydrologic cycle and watersheds are also highly dependent on soil conditions which determine the rate of percolation, surface cover (soil, vegetation, paving), and climatic factors.

Discussion Questions:

1. What surface and subsurface sources could result in contamination of groundwater?

2. When groundwater is contaminated, what are 3 consequences to humans and/or animals?

3. What are three ways to prevent groundwater contamination?

4. Once an aquifer is contaminated, how long do you think it would take to cleanse itself? What could be done to clean it?

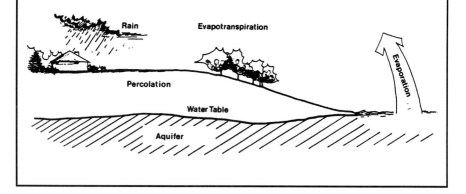

A Case Study: The Rouge River Education Program

Introduction

Many educators seek an educational framework in which to apply the watershed analysis measures described in earlier chapters. As a case study, the Rouge River Education Program provides a good and widely-disseminated model of school-based watershed education.

This chapter describes the development of this program as well as new directions for watershed education programs. There are many fine watershed education programs around the world. The Rouge River Education Program is offered here because it was very influential in the early development of the Global Rivers Environmental Education Network (GREEN). But it only represents one model—many more have emerged within GREEN. Each educational model reflects the aspirations, experiences, and talents of its participants.

The goal of the Rouge River Education Program is to develop a citizenry in the Rouge River basin that is aware and concerned about the river. Objectives of the program include:

- ➤ linking diverse schools and communities together—rural, suburban, and city through the common thread of the Rouge River;
- ➤ providing a watershed focus and watershed-wide analysis;
- ➤ increasing student problem-solving skills;
- ➤ providing an interdisciplinary focus (science, language arts, mathematics, visual arts, social studies); and,
- ➤ promoting student empowerment and action-taking.

There are four cornerstones to the Program that support these objectives: **watershed analysis** (water quality testing, benthic macroinvertebrate sampling, remote sensing, land use surveys); **telecommunication** (Internet);

school-community partnerships; and, **educational research.** Elementary schools, middle schools, and high schools throughout the Rouge River watershed are engaged in this program.

The Rouge River

The Rouge River is a part of the Great Lakes-St. Lawrence drainage basin. It is only one of hundreds of river systems draining into the Great Lakes, eventually flowing through the St. Lawrence River to the Atlantic Ocean.

The Rouge River, in its lower reaches, is considered by many local people to be an eyesore. Few realize that the headwaters of this 465 square mile watershed begin in rolling countryside. It is only when the river flows into metropolitan Detroit that the effects of the 1.5 million people who live in the watershed become truly apparent.

About 60 percent of the watershed is urbanized. In these areas, rain and melting snow run off impermeable surfaces like roads, parking lots, and rooftops. This runoff carries salts, dirt, oil, grease, heavy metals, and chemicals into nearby sewers and watercourses.

Underlying much of this urban area is an antiquated sewer system that carries both stormwater runoff and sanitary wastes from homes, schools, and businesses. The present capacity of this combined sewer system is inadequate to handle the increasing volume of urban runoff. As a result, untreated sewage is sometimes shunted into the Rouge River to avoid overwhelming the system.

Untreated sewage, unstable river flow, and the naturally slow flow of the river have contributed to its degradation. The major environmental problems facing the Rouge today—thick mats of sewage blanketing the bottom, fluctuating oxygen levels, contaminated sediments, undercut banks and massive log jams—have become the catalyst for educational challenges and citizen action.

Beginnings of the Program

The Rouge River Education Program was formed in response to a growing realization that the river had been abused and neglected for too long. Friends of the Rouge, a nonprofit community organization, initially sponsored this program in 1987 to encourage stewardship and restoration of the Rouge River.

Students and educators from the University of Michigan's School of Natural Resources worked closely with Friends of the Rouge to implement this program in 16 high schools the first year. Teachers in these schools were supported by University students who were trained in computer net-

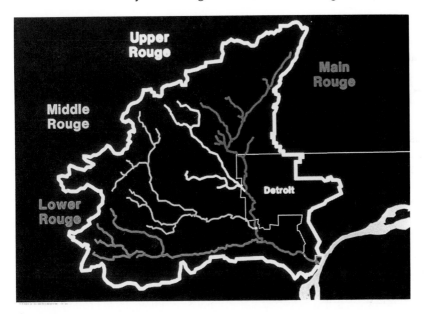

Figure 8.1. The Rouge River watershed and its four branches—Main, Upper, Middle, and Lower—that empty into the Detroit River.

working and water quality testing; these students assisted teachers in the classroom during the two week program. To guide the program in its early stages, the following steps were taken:

1. An advisory committee was formed of science and social studies curriculum coordinators, local science teachers, water resource professionals, and representatives from Friends of the Rouge. From this group, teachers were recommended for the first year of the program.

2. Criteria were established to select interested teachers. The two criteria were: a commitment by schools to support the program, and a commitment to build a diverse program—balanced among suburban and city schools.

3. A Letter of Agreement was created with a description of the program. The letter included: responsibilities of participating teachers and schools (time and resources), as well as a commitment of resources and support by Friends of the Rouge. The principal, teacher, curriculum coordinator, and Friends of the Rouge signed off on this agreement.

Figure 8.2. The advisory committee was instrumental in helping to plan the project and in carrying out an on-going evaluation.

Program Planning Session

The objectives of this session were to begin building a network through personal contacts among teachers, and between teachers and university personnel; and, to involve teachers from the beginning in program planning. Teachers were given an overview of the program and a sense of how it might work at the classroom level. A special effort was made to solicit feedback from teachers on a curricular framework. Their ideas and concerns were translated into a program that was both innovative and pragmatic. Teachers specifically commented upon, or helped shape:

➤ the structure of future workshops;

➤ institutional realities within their schools that could hinder the program;

➤ the importance of a reliable, accurate, and safe watershed analysis approach; and,

➤ the need for technical assistance in water quality measurements, and in computer networking.

The selection and planning process helped to build an educational program known as the Rouge Model. There are several characteristics of this model: a watershed focus; schools connected by a computer network; a collaborative learning approach that includes teachers, students, and

Figure 8.3. Teachers reviewing the procedures for running the water quality tests.

researchers; a shared monitoring day; and, an intense two week watershed education program that culminates in a Student Congress.

Over the years, the model has remained largely intact in the Rouge. Watershed education programs are, however, adapting and growing in response to: reform efforts in education, a movement towards building sustainable communities, and an ever shrinking pool of government funds.

The Program Today

A decade later, the Rouge River Education Program now numbers 100 high schools, middle schools, and elementary schools scattered throughout the Detroit metropolitan area. The program is one element in a much broader effort to clean up the Rouge River, known as the "Rouge River National Wet Weather Demonstration Project". The EPA has funded this project to monitor, model, and mitigate the effects of wet weather flows on the Rouge River. The Rouge River Education Program is the most visible public education and involvement portion of this Demonstration Project Friends of the Rouge still administers the education program, and the staff includes a full-time director, a full-time coordinator, a curriculum consultant, and a telecommunication consultant.

A series of workshops form the backbone of this Program leading to a 2–3 week intensive watershed education unit and culminated by a Student Congress.

Assessment Workshop

How do we measure success in these programs? Educational and program assessment is critical for three reasons: it provides information that can help improve the program, it offers a concrete way of demonstrating success for foundations and other funders, and it helps build more credibility for the program within the school curricula.

Watershed education programs in the northwestern United States (like the Nisqually and Budd-Deschutes River Programs) have led the way in designing an assessment framework that considers four categories: students, teachers, community, and environment. The Rouge Program has developed an Assessment Guide that reflects the Michigan Essential Goals and Objectives for Science Education K-12 and the Michigan Educational Assessment Program (MEAP) and proficiency tests.

The REP Assessment Guide utilizes *constructed-response* tasks about water quality *investigations,* that require students to develop or convey their understanding of the scientific process, and *text critique* questions that encourage the synthesis of information gathered from real newspaper articles about the Rouge River.

Telecommunication Workshop

Teachers, computer coordinators, and a student are invited from each participating school to a telecommunication workshop. Friends of the Rouge along with host sites offer four regional telecommunication workshops. Most of the host sites are either well-equipped high school computer labs, or are university computer labs.

These workshops are structured to offer people guided practice in the use of the Internet. Topics include: how to navigate the Internet, how to input water quality data, how to upload and download, and how to connect the REP web site. Future telecommunication workshops will include the use of *Riverbank,* an on-line database developed by the University of Michigan and GREEN and how to design host pages for their school on the Internet.

Bus Tour

In mid-April, participating teachers are invited to take a bus tour of one section of the Rouge River. The tour is valuable, particularly for new teachers to the program, because it offers an integrative approach to watershed education. The tour begins in a relatively upper middle class area of the watershed and ends in a low-income neighborhood near the river's mouth. Social implications of both water pollution from combined sewers, and air pollution from industries near the mouth of the river take on real significance on this tour. Industrial development and the growth of the auto

Figure 8.4. Teachers traveled by bus to sites along the downriver portions of the Rouge River to observe the influence of water quality on the people living in the area.

industry, the historic struggles of unions, the history of ethnic enclaves, and the evolution of wastewater treatment are all subjects that come alive through this tour. Some teachers arrange to take their class on the same guided tour later in the school year.

Remote Sensing Workshop

As noted in chapter 7, the two main forms of remotely sensed data are aerial photography and satellite imagery. Aerial photographs are ordered to include the area of the schools represented at the workshop. Teachers begin with the use of topographic maps to find latitude and longitude of a site. Aerial photographs are used to begin to identify landforms. Satellite images like Landsat Thematic Mapper false-color images are used to interpret land uses in the watershed. Future workshops will include hands-on work in a Geographic Information System (GIS) called ArcView II, and with Multispec, an imaging software that allows the manipulation of color bands to highlight specific land uses in satellite images.

The Watershed Analysis Program

Day 1: Orientation to the watershed (Monday)

The first day of the program is an orientation session. Students reflect on how water quality affects them personally and their community. They view a slide-tape presentation and videotape on the Rouge River, discuss the boundaries of the watershed and pollution issues, and are given an overview of the 15-day program.

Some schools choose to begin with the history of the Detroit area and the historical role of the river for transportation. Other schools collect benthic macroinvertebrates to place in aquaria for observation.

A resource person from the community or from the University assists in each new classroom. Some schools sign onto the computer network and describe their school and their proximity to the river.

Days 2 and 3: Learning to Analyze the Watershed (Tuesday and Wednesday)

These are integral practice days that enable students to practice physical, biological, and chemical analysis of the river and its watershed. Students learn the meaning of the water quality tests and how to perform them safely and accurately. Students learn to identify common benthic macroinvertebrates and how to derive simple indices. Students also practice how to conduct a physical survey of the river. Depending on the teacher and the level of the class, each student either practices and masters a single measure, or is exposed to all of the measures but only made responsible for one. A teacher or a resource person is present at each of the stations to facilitate hands-on work with the test kits.

Besides reviewing the test procedures, students also learn how specific tests relate to other tests within the Index, and the significance of these measurements to aquatic life.

Day 4: Monitoring Day (Thursday)

This is the day to collect data. Each school travels to its assigned monitoring site where secondary students conduct dissolved oxygen, fecal coliform, pH, temperature, total phosphate, and nitrate tests. Water samples are collected for later use in measuring turbidity and total solids.

The biochemical oxygen demand (BOD) sample is taken five days earlier, allowing this test to be run as well. However, an additional BOD sample is taken on this day to be tested five days later. Turbidity is measured with a turbidimeter. Total solids and fecal coliform samples are delivered to measuring scales and incubators set up around the watershed.

Elementary students run dissolved oxygen, pH, and temperature. Elementary and secondary students sample for benthic macroinvertebrates, conduct stream surveys, and measure velocity and discharge.

Safety is a major concern. All students who sample or conduct tests wear rubber gloves and safety goggles as a safeguard against possible infection. Samples are taken with a homemade sampling device that holds a sampling bottle. These devices, built by either the school's shop department or by the class, enabled the students to avoid contact with the river and possible contamination. Finally, everyone is requested to wash their hands upon completing the tests.

Figure 8.5. Early classroom discussions often focused on the Native Americans that lived along the Rouge River 3,000 years ago.

Day 5: Determining the Water quality Index (Friday)

With all of the data assembled, students convert it into Q-values and record it on the chalkboard. From this data, the class derives a Water Quality Index (WQI) for their monitoring site. Some schools begin graphing their data and Index to share with other schools.

This is the stage when the computer network becomes more important. Students enter the raw water quality data and their Water Quality Index into the computer, making this information accessible to students at all of the other schools. In addition, students question each other via the computer on the significance of the WQI and of specific measurements in terms of public health, especially the fecal coliform data.

Days 6 and 7: Exchanging Information and Computer Graphing (Monday and Tuesday)

On the sixth day, some students download data from other schools from the web site REP, and then, using computer graphing programs such as Claris Works, Cricket, or Excel, they graph each water quality parameter and the WQIs along their branch of the river. When the WQI values are graphed, students can easily identify changes in water quality along a given stretch of the river, as

Figure 8.6. Most of the schools were within walking distance of their monitoring sites.

well as water quality problem areas. The computer network becomes a central focus as students seek answers to discrepancies in the results from location to location.

In addition to the Rouge River computer network, students can communicate with individuals from other watersheds through an international network organized by GREEN. Students and teachers can exchange information regarding water related problems with their counterparts in watersheds around the world. (See Chapter 9 for more information about GREEN.)

While science students are graphing the water quality data, social studies students are engaged in a role playing simulation designed to help understand the implications of different land use decisions on water quality within a watershed.

Day 8: Building Skills to Identify and Define Problems and Their Sources (Wednesday)

On this day, students may participate in a brainstorming activity that focuses on possible water quality problems indicated by the data. Each class then comes to an agreement on a specific problem it wants to address.

For example, students at one high school (North Farmington) noticed that fecal coliform levels in a tributary of the Rouge were significantly higher below a pipe than above it, leading them to suspect that the pipe was discharging sewage. Further investigation revealed that raw sewage was, in fact, entering the stream at that point from a malfunctioning pumping station. The problem was corrected promptly after bringing it to the attention of the City Engineer.

Day 9: Building Skills to Utilize Information (Thursday)

The students identify water quality issues that they want to know more about, and to develop research skills for finding the information they need . Telephone books, phone calls, libraries, parents, and even the computer network helps them find key information, such as the ownership of local developments that are contributing to soil erosion, and the management of drains upstream from their monitoring site.

This line of questioning also leads to efforts to solve problems. As a classroom activity, students decide what they would like to do about the problem, how they would like to see it resolved, what resources are available to them, which individuals stand as potential obstacles to solving the problem, and which actions can be taken individually and as a group.

Figure 8.7. EcoNet was also used to establish direct communication between schools and community leaders in different watersheds around the world.

Day 10: Organizing Information and Discussing Action Strategies (Friday)

This day is devoted to helping class representatives prepare for the Rouge River Student Congress to be held the following day. At the Congress, student representatives are expected to present their watershed analysis data with a thorough understanding of potential causes. Potential water quality problems are verified by communicating with schools upstream by computer. In addition, students develop an action strategy based on their goals and resources to share with other schools at the Student Congress.

Day 11: Rouge River Student Congress (Saturday)

On Saturday of the second week, nearly 270 students, teachers, resource people, administrators, and water resource professionals gather at one of the participating schools in Detroit for the Rouge River Student Congress. Students come prepared to share their water quality data with other students, to talk about water quality problems in their communities, and to generate recommendations for solving these problems along each branch of the river. They take part in skill-building sessions to help them act on their understandings and commitments.

Detroit Free Press

Section A, Page 3   SECOND FRONT PAGE   Sunday, May 17, 1987 •

Human wastes found in Rouge

Effluent in river despite dry period

By JOEL THURTELL
Free Press Staff Writer

When students from 16 high schools in Wayne and Oakland counties tested Rouge River water earlier this month they found enough old tires, junked cars, shopping carts and empty beer bottles to stock a town dump.

They also found the unexpected — human waste flowing into the Rouge from a sewer in Farmington Hills even though it was dry weather.

"We found that our river is as bad as we thought it would be. We found no living organisms," wrote River Rouge High School student Malika Noland. "There were oil slicks floating on top of the water, and garbage. The water was dark green and smelled very bad."

Effluent entering the river during rainy periods may be disgusting, but that's how the system was designed in the 1940s and 1950s — to let storm drains conduct sewage to the river at times of peak water flow.

But the dry weather sewage measurements by North Farmington High School students on May 7 shocked James Murray, the Washtenaw County drain commissioner who also is chairman of the State Water Resources Commission and president of Friends of the Rouge, a private group dedicated to improving water quality in the Rouge.

"It tells you the CSO (combined sewer overflow) has got crap going out of it, and that's illegal during dry water flow," Murray said.

HE DISCUSSED implications of the water test results Saturday at Detroit's Redford High School with about 50 students from the 16 Rouge basin high schools participating in the interactive Rouge River Water Quality Project sponsored by Friends of the Rouge.

But even before Saturday's meeting, students from North Farmington High wrote in the project's computer bulletin board that "we are having the city engineer of the city of Farmington Hills come to our class May 28 for us to learn of the the present system here and for us to share our data with him. Our CSO was still draining days after the last rain."

Erin English, a 16-year-old sophomore biology student from Redford Union High School, described her efforts at drawing water samples from the Rouge's upper branch: "It was really gross — we saw a couple of tires in the water. We didn't see any fish."

Data from the 16 student groups showed consistently increasing degradation along the river, said William Stapp, a University of Michigan natural resources professor who directed the project.

The Rouge runs from spindly streams merging in Oakland County to its mouth near Zug Island in River Rouge. More than 1.5 million people live in the Rouge River basin.

Sewage was well below levels considered dangerous for swimming in Novi, Birmingham and Plymouth, but fecal coliform levels increase steadily as the river flows downstream. Near the river's mouth, a student group measured fecal coliform 127 times higher than the maximum level allowed for swimming and 25 times higher than the level allowed for boating. The levels were 25,400 times higher than the maximum level allowed for

Figure 8.8. North Farmington students located a cracked sewer line leaking wastes into the Rouge River at their monitoring site. Reprinted by permission of the *Detroit Free Press*, May 17, 1987.

The day is organized into morning and afternoon sessions. In the morning, the focus is on the sharing of water quality data and student action plans. Students divide themselves into discussion groups based on the branch of the river monitored by their schools. Topics of discussion range from land uses, to the behavior of individuals, to economic factors that contribute to declining water quality.

Out of these morning work sessions came the following recommendations:

- To restrict development in sensitive areas, particularly wetland and floodplain areas near the Rouge River and its tributaries;
- To encourage increased media coverage of the Rouge River;
- To begin recycling programs within schools and the community, including motor oil, aluminum, newspapers, and other materials, in conjunction with a reduction in the use of plastics and styrofoam;
- To raise awareness through education of younger students by placing posters around the school about the Rouge River and by conducting a community survey and education program.

Also in the morning, teachers hold their own session to review the program. Resource people are briefed about the afternoon workshops. Students and water resource professionals have the opportunity to ask questions of each other before the entire group. During lunch, students sign up for an afternoon skill-building workshop to help them address a specific set of problems.

The focus of the afternoon is on skill-building workshops to help students take action. Each workshop is led by several facilitators with specific expertise. The workshops include:

- Radio/public service announcements (given by a student)
- Video production
- Community/school organizing
- Editorial writing
- Public hearing/remedial action plan
- Artwork/posters
- Communicating with legislators/local officials
- Street theatre.

The purpose of these workshops is to expose students to a specific skill area so that the basic foundation for taking appropriate action is in place when they return to their schools. At each workshop, students are introduced to the skills needed and then asked to create their own project, whether it is a poster drawn, a letter to the editor written, or street theatre performed.

All of the students reconvene after working in groups or workshops to share what was learned.

The concluding session of the day is a demonstration by the students of the skills they had learned in their afternoon workshops. Students articulate their commitment to the clean-up of the Rouge River through a range of media—from the spoken word to artistic creations. By the end of the day, students are anxious to report back to their classmates on what has occurred at the Congress and on some specific actions that they could take as a class.

For many classes, this is the end of the program. However, some teachers are able to continue the project for several more days, giving students more time to achieve their action goals. This is ideal because it allows students to use new skills from the Congress.

Days 12-14: Individual and Group Action (Monday, Tuesday, and Wednesday).

Representatives to the Student Congress report back to their classes on their experience at the Congress and the skills they think can be focussed on problems relevant to the class.

The Social Studies Game

To add a social studies component to the program, an interactive role-playing simulation game has been developed and played in a number of different schools in the Rouge watershed. The game introduces students to some of the issues inherent in making land-use decisions by having them play the characters who make those decisions. It is a fascinating experience for students, and it has been very successful in educating students about the multidisciplinary challenges of natural resource management.

Some classes continued their involvement with the river by participating in the Rouge Rescue, which is organized by Friends of the Rouge and held in early summer. The students join with other citizens to pull debris and garbage out of the water. To publicize the Rouge Rescue within their own schools, students use their public address systems, write articles for their school newspapers, produce skits for school assemblies and talent shows, and paint posters urging students to help with the cleanup.

Some classes reflect on what they or their parents were doing at home—individual behaviors that can be eliminated or changed, such as not using lawn fertilizers and pesticides, or limiting the amount of household chemical cleansers used. Some made a commitment to ride bicycles more often.

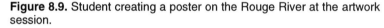
Figure 8.9. Student creating a poster on the Rouge River at the artwork session.

Students pursue other positive ways to help the environment. Some decide to voice their opinions on a proposed development at a city council meeting. Others chose to start recycling aluminum, newspapers, and motor oil instead of throwing these in the garbage or down the sewer. Some students decide to continue monitoring for specific parameters to keep attention focused on problems they had discovered previously.

This case study demonstrates the potential of education when it is focused on a community problem—one that impacts students' lives. In just two weeks the young people in this project moved from knowing little about the Rouge River to a deep understanding of the Rouge watershed and its problems. The project gave students the freedom and opportunity to explore problems in their community.

Taking advantage of this opportunity gave students, in turn a sense of accomplishment and empowerment—a feeling that they could, in fact, make a difference. That feeling permeated the Rouge River Student Congress and was felt by teachers as well students.

Figure 8.10. Students from multiple schools discuss water quality results.

The result is that many students act on the information and the understanding that they hold—a good start towards building the citizenry needed to solve the complex problems of our world.

Figure 8.11. Some students gave strong consideration to alternative strategies for lawn care at home and at their schools.

Examples of Student Actions

Students find some very creative ways to educate their communities about water quality. In the Pacific Northwest of the United States, groups of students stencil messages on curbs to warn people against dumping wastes into storm drains (see Chapter 7). In Michigan, students persuade city officials to post a sign warning against swimming in the river because of high fecal coliform counts.

In other areas, students help to revegetate river banks, and they attend hearings to oppose poorly planned development projects. They work with school officials to limit the amount of fertilizers applied on school grounds, and they helped to develop hazardous waste policies for their schools. Some students create videos about the river as class projects.

Students have undertaken many activities outside the classroom. Their studies increase not only their own awareness and concern for water quality, but also that of their families, neighborhoods, and communities. The protection of some rivers can be traced back to the work of local students.

Innovative Watershed Education Programs

Since 1987 and the Rouge River Education Program hundreds of school-based watershed education programs have emerged. Now there are literally dozens of models for how to conduct a watershed education program. Different models evolve out of different physical surroundings, and from leaders with diverse backgrounds and experiences. Here is a sampling of innovative watershed education and stewardship programs.

Interdisciplinary Programs

The Saginaw River is one of the best models of an interdisciplinary K–12 watershed education program. Students produce videos, create poetry, write songs about the Saginaw River, and author books about the history of the river.

Another wonderful example of an integrative watershed program comes from the Delaware River Program. In this program, two teachers have together constructed an entire year of middle school around the theme of a watershed. Their students build canoes and paddle them down the river, they build models of the wastewater treatment plant, they study geography and its connection to history, and they develop a puppet show about the river (to name just a few of the activities).

Community Support

The Olympic School District and the Nisqually and Budd-Deschutes River Programs have done some very innovative work in building strong community support for their programs and in developing innovative assessment strategies.

Cross-Cultural Program

The Project del Rio on the Rio Grande/Bravo along the United States and Mexico border is an example of a strong cross-cultural program. In addition, it has led the way in developing a quality assurance and quality control program for their monitoring program.

Student Action

Lee County School District in Florida is one of the best examples of students engaged in local environmental issues. The students have protected bald eagle nesting sites, and stopped unnecessary development in that part of Florida.

Community Service

In Sydney, Australia there is an innovative program called "Kids, Companies, and Creeks" in which young people perform environmental audits for local businesses, such as monitoring the effluent or giving advice about preventing runoff or erosion.

Anoka, Minnesota (Anoka High School) engages high school students in monitoring local rivers in tenth grade, and as seniors participating in a "Students in Government" Program. Students participate in three community task forces responsible for looking at environmental health, community health, and business health of their community.

As the global network of watershed education programs becomes even more extensive and diverse, it will be important to share successful models so that all programs may benefit.

International Rivers

Cross-Cultural Opportunities

What can we gain by taking a global approach in our studies of water quality issues?

1. We can learn about the uses and water quality concerns of people all over the world.
2. We can learn about ways of thinking, feeling and responding to water quality and other environmental issues.
3. We can learn about alternative ways of solving common water quality problems.
4. We can encourage international cooperation through mutual respect, beginning at a local level.

It is easy to understand the similarities we share with others when we consider that all life depends on water. Life evolved in water, and no living organism can survive without it. Not surprisingly, people throughout the world are faced with similar challenges of finding potable water, storing water and transporting water.

Historically, we can find many examples of cultures solving water-related problems in similar ways. For example, irrigation systems were part of the Mesopotamian, Roman, and Mayan civilizations. Water was used to measure time in China and Egypt.

The Egyptians invented a 365-day calendar based on the annual flooding of the Nile.

Birch bark canoes were used for water travel by North American Indians, dug-out log canoes by Africans, rafts of balsa logs in Peru, outrigger canoes in the Pacific, skin kayaks in the Arctic, and junks in China. These forms of river transportation enabled cultures to explore their surroundings, and trade with other cultures. More importantly, river travel promoted the exchange of ideas and technologies among cultures.

Figure 9.1. This program provided the opportunity for students to explore perceptions of people located along different rivers in the world and in different cultural settings.

Land use patterns today reflect our dependency on water. Most major cities are located near rivers or ocean bays. Industries, also dependent on water, are frequently located near water. Ironically, the same water used for drinking supplies is also often used for the disposal of wastes. Despite all of civilization's "advancements," we are all still tied into the natural cycles of water on this planet. Monsoons, droughts, groundwater replenishment rates and the self-cleansing abilities of water bodies are phenomena that we cannot successfully manipulate.

The relationship each of us feels with the earth influences the way we treat our environment, the laws we make, the religions we practice, and the alternatives we choose in meeting our water needs. For instance, many Americans believe humans are the only creatures in nature that have souls.

Figure 9.2. Students in this program had the opportunity to communicate through computer conferencing with students in Germany regarding land use practices along the Rhine River.

The rest of nature is viewed as a source of "resources" to serve human purposes. Related to this attitude is the belief that progress through the control of nature and personal improvement is possible. These attitudes are reflected in our water regulations and technology. Water quality standards are determined by how we want to "use" the river (e.g. industrial, recreational). And Americans have used their technical "ingenuity" to harness the energy of rivers with big dams that have radically changed the surrounding environment.

In contrast, some cultures believe that humans are not intrinsically more important than other living things found on the earth. Hindus in India, for example, believe that life is an endless cycle in which souls can assume an infinite number of forms. Hence, their relationship with the earth is quite different. The Ganga (or Ganges) River in India plays an important role in the Hindu religion. Hindus strive to wash in the river's water at least once in their lives and have their ashes scattered in the river when they die. Water taken directly from the river is considered holy and used in religious ceremonies; once treated, it is thought to lose its spiritual value. At the same time, however, urban sewage is flushed into the river through ancient stone drainage systems. This is not seen as a serious problem since, according to the Hindu faith, humans do not have the power to pollute the Ganga.

Figure 9.3. One-fourth of Australia's population lives in Sydney, which has one of the most attractive recreational harbors in the world.

Attitudes towards nature are deeply rooted and are formed by cultures as adaptations for living on this planet. We evaluate ourselves and others by our standards of what is normal, not by the standards of other cultures. An American, for example, might judge India as being underdeveloped or unproductive, while someone from India might view Americans as irreverent and unspiritual. The culture we are raised in influences us to see the world differently.

No one culture's attitude toward nature is necessarily better than another's. In fact, our cultural differences can be an asset. Different cultures focus on different aspects of nature. The combination of all of our perspectives provides us with alternative ways of looking at the world, and solutions to common problems that we might not have considered on our own. By being sensitive to cultural differences, we can get the most benefit from this opportunity.

People using this manual are linked in two ways. First, we all have a concern for water quality. More importantly, we are all connected because of the global nature of the water cycle. Water issues can never be just local. The decisions we make locally—our laws, land use patterns and technology—have an impact on the rest of life on earth in ways that we don't always realize. Working together with other cultures can make us aware of our responsibilities as citizens of the world.

Figure 9.4. Students in India are examining the quality of their rivers and sharing this information with the Rouge River participants.

Rivers that form national boundaries, such as the Rio Grande, or flow through more than one country, such as the Nile and Rhine, provide ideal opportunities to work with others on water quality issues. Exchanges made with other countries which don't share rivers can also be rewarding. Using pen pals, phone calls, FAX, travel, or a computer-based message system are some of the ways countries can communicate.

The Global Rivers Environmental Education Network (GREEN)

What is GREEN?

"Globalization" is a word you commonly hear in educational circles these days. It refers to the widely acknowledged need of schools to equip their students for an interdependent world, linked by a closely coupled world economy. This shrinking world is brought closer together by massive environmental problems and issues that transcend national and even continental boundaries—issues that we can address only through an unprecedented degree of global cooperation.

One major challenge that will increasingly confront environmental educators is to develop curricula and instructional strategies that emphasize the global aspect of local environmental issues but do not overwhelm the

Figure 9.5. GREEN participants discussing the involvement of Taiwan governmental officials and secondary students with the Director of the Environmental Protection Agency.

students or cause them to lose hope. How can we educate and empower students to take action on local issues, while simultaneously developing within them a global, cross-cultural perspective on these issues? How can we best encourage this first generation of truly planetary citizens to assume responsibility for their shared, imperiled home?

One promising approach to meeting this challenge is the Global Rivers Environmental Education Network (GREEN), an international network that seeks to bring secondary school students, teachers, and communities around the world closer together through the bond of studying and improving our common river systems. GREEN was initiated by Professor William B. Stapp and graduate students of the University of Michigan's School of Natural Resources in 1989. In 1999 GREEN joined forces with Earth Force, a national non-profit organization committed to young people changing their communities while developing life-long habits of active citizenship. GREEN has developed into a multifaceted, global communication system that invites participants to examine ways that land and water usage and cultural patterns influence river systems and visa versa. The network encourages learners to become involved in complex, real-world concerns that extend across traditional boundaries.

In this way, GREEN works to achieve three interrelated goals:

➤ It acquaints students with the environmental problems and characteristics of their local watershed, giving them "hands on" experience in chemical, biological, and sociological research.

➤ It empowers students through community problem-solving strategies, thereby enabling them to see the relevance of subjects they learn in school to the "real world."

➤ It promotes intercultural communication and understanding, and thereby fosters awareness of the global context of local environmental issues, and of the significance of cultural perspectives in choosing effective problem-solving strategies.

Why Rivers?

Rivers were chosen as the central focus of the project primarily because they are a reliable and informative index of the environmental quality of their watersheds. But rivers also form a nexus for relating chemistry to biology, and for relating the physical sciences to the social sciences and humanities. Rivers bind together the natural and human environment from the mountains to the sea, and from farmland to the inner city. In fact, 85 percent of the world's human population lives on or near a river. For these reasons, the study of rivers forms a coherent curricular framework for the study of a wide range of environmental issues. Rivers also contain a historical perspective on cultures and society, forming an ideal basis for learning about cultural diversity and engaging in cross-cultural dialogue.

Through involvement in a network on local rivers, students share information, techniques and different approaches to problem-solving. They also learn that their investigations are valued by their peers elsewhere in the world. Students are motivated to further their understanding of their watersheds and to work to resolve some of the water quality problems they have discovered. GREEN is therefore a program designed to bring individuals closer together and encourage them to develop a sense of responsibility for their communities and their planet simultaneously.

The History of GREEN

Under the guidance of William B. Stapp, a committee of university students with backgrounds in environmental education and international issues was organized as an advanced level Environmental Education class in January of 1989. The committee established the vision and goals of

Figure 9.6. Swaziland villagers washing clothes in water of poor quality. Students are creating an alternative site for washing clothes with groundwater as the source.

GREEN as they laid the foundation for this international network. The committee organized and facilitated 22 workshops in 18 nations in Africa, Latin America, Europe, Asia (including the Middle East), and Australia during the summer of 1989. The workshops brought together educators, administrators, students, citizens, resource specialists, and representatives from governmental and non-governmental organizations to exchange ideas on watershed programs. One of the aims of the workshops was to discuss approaches to experiential, interdisciplinary environmental education, and to explore how water study programs could support the educational goals of each nation.

The workshops were fruitful in stimulating ideas and plans highlighted by the following points:

1. *Water quality varies.*
   Rivers varied in quality from very pristine to highly polluted rivers that contained at mid-day: 0.0 ppm dissolved oxygen, 1.2 million colonies of fecal coliform per 100 mL of water, 138 ppm biochemical oxygen demand, and high concentrations of heavy metals and toxic organics. Nitrogen levels in some estuaries had increased by 200 percent since the 1950s. One bay has received 600 tons of inorganic mercury since 1953.

2. *School structure is a factor.*

School curricula vary greatly throughout the world. Some nations are very decentralized and flexible in permitting interdisciplinary water quality programs, while other nations have highly centralized systems where students are prepared for rigorous national examinations. However, one such nation permits students to substitute an independent project, such as an individual river study, for the national biology exam.

3. *Participants respond enthusiastically.*

There was great interest among the participants in water monitoring programs that enabled students to link education to real-life experiences; work between disciplines; share information through computer networking; and take action on information collected. Some participants expressed the concern that science education should remain value-free; that leaving school grounds would not be permitted; and that water monitoring kits and computers were too costly.

4. *Interest and experience in using computers varies.*

Some schools used powerful computer networks to share information on water programs and access national data bases on rivers. These schools were interested in making better use of international computer networking opportunities. Teachers in some other schools viewed computers with less enthusiasm, believing that they would overshadow less high-tech activities and would be more costly, time-consuming and impractical.

The Creation of the Infrastructure and Networking

Many of the nations that hosted or attended workshops that first summer initiated their own school-based water monitoring programs. For example, Taiwan initiated two programs monitoring rivers; schools in Germany created an environmental monitoring network; and Israel is incorporating river studies into their national Senior High School Curriculum. Several countries appointed GREEN Country Coordinators to oversee GREEN activities and prepare articles and other educational materials. In addition, funding from private and public sectors was allocated to develop programs and obtain equipment and supplies.

With so much interest shown in learning more about student water monitoring and networking, GREEN focused on creating a globally accessible network and on the development and dissemination of relevant

Figure 9.7. Two teachers testing the nitrate content of water from Lake Nakuru, Kenya.

resources to participants. GREEN began to publish a newsletter; develop curriculum guides for various age levels; conduct teacher training workshops; and support other national and international river education projects, such as on the Rouge River in Detroit and the Rio Grande.

The Expansion of GREEN

In 1993 GREEN became a private, not-for-profit organization whose mission is to improve education through a global network that promotes watershed stewardship. GREEN's strategies for achieving this goal includes providing program support, enhancing information exchange, developing cross-cultural opportunities, encouraging educational reform, and conducting educational research.

Worldwide response to GREEN has been truly phenomenal. Internationally, GREEN has grown to reach thousands of students involved in watershed projects in countries as disparate as Bangladesh, the Czech Republic, and Argentina. There are now GREEN Country Coordinators in 52 nations and active programs in 85 countries. Some nations, such as Germany and Australia, have established extensive national networks to facilitate communication.

Within the U.S., 35 states have developed watershed-wide programs and many other schools are monitoring independently. The Teacher Enhancement Program, funded by the National Science Foundation, allowed teachers in five watershed programs in the United States to work together to further develop the GREEN watershed education model and to create plans to disseminate the model to other interested schools in their regions. GREEN has also worked with Native American educators to help assess their environmental education needs and to provide program support.

The development of GREEN can be largely attributed to a sound environmental education model and to a vision of transcending cultural boundaries with global environmental issues. In order to better support watershed programs throughout the developing world, GREEN in cooperation with the LaMotte Company, developed a Low-Cost Water Monitoring Kit. This kit allows users to test for eight of the nine water quality tests detailed in this manual.

Components of the GREEN Network

Using communication to effect environmental change and empower students is the backbone of GREEN's philosophy. The network presently disseminates a quarterly newsletter to educators, ministry officials and other resource persons in over 130 countries. A series of GREEN International Computer Conferences have been established, and GREEN has created the Partner Watershed Program to enable schools in different nations to share information. GREEN also serves as a clearinghouse and a resource for watershed education programs.

1. *The GREEN Newsletter*
   The GREEN Newsletter is the most extensive and important communication tool of the Network due to the accessibility of the mail system and the ability of the newsletter to impart information and foster the GREEN spirit. Examples of topics of newsletter articles include: ideas for starting a water monitoring program, low-cost monitoring techniques, and models for student action taking and community involvement. Each issue highlights exciting programs that serve as examples of local watershed education programs around the globe.

2. *The GREEN International Computer Conferences*
   The GREEN Conferences are international, electronic forums for the exchange of student-collected water quality data, reports from watersheds, solicitation of watershed partnerships, and the ideas and concerns of students, teachers and other professionals engaged

in GREEN programs. Participants are able to communicate their experiences and receive almost instantaneous responses from around the world.

In addition, individual watershed programs use on-line networks to host local and regional computer conferences. These conferences allow the students to enter their data and communicate interactively with other schools in their watershed.

The GREEN International Computer Conferences, and many of the watershed-specific computer conferences, are hosted by the Association for Progressive Communication (APC) networks, an international coalition of independently operated computer networks in 18 countries that together extend their services to more than 133 countries. Via EcoNet in the United States and its APC counterparts, the GREEN conferences can be shared with regional, state, and local educational telecommunications networks around the world.

The GREEN office in Ann Arbor, Michigan, may be contacted for details via Internet at <green@green.org>. Information about GREEN is also available via Internet Gopher at <gopher.igc.apc.org> in the Education & Your/Projects menu, or via World Wide Web at <gopher://gopher.igc.apc.org/00/orgs/green>.

3. *The Cross Cultural Partners Watershed Program*
The Cross Cultural Partners Watershed Program has sparked remarkable interest among schools worldwide. Through the program, GREEN matches schools involved in watershed projects in different countries to further enhance cross-cultural sharing. The pilot student-to-student links, initiated in 1991, consisted of schools in the United States, Canada, Mexico, Hungary, Australia, New Zealand, and Taiwan. Since then, many other partnerships have been established. Using local water quality issues as a medium for discussion, students exchange personal cultural perspectives and concerns for the environment, along with ideas for improving their environment. Generally students communicate by mail and computer to share their thoughts.

The aim of this cultural exchange program is to stimulate greater international awareness in students while at the same time motivate students to work toward improving their local waterways. GREEN is developing materials that will help partnered schools achieve this goal.

Figure 9.8. Native American youth analyzing water quality information.

Conclusion

The concept of student environmental monitoring is an exciting one. The prospect of students, teachers, researchers and other professionals communicating about their local environment concerns nationally and internationally is significant. It is GREEN's vision to reach this potential so that students can be empowered to become active learners and problem-solvers through a successful networking system incorporated into their educational process.

Figure 9.9. Australian students using a new nitrate test kit developed by Tintometer, Inc. (United Kingdom) for testing in saline waters.

Through the broad response that GREEN has received, it is clear that educational systems around the world are ready to incorporate real-world topics into their classrooms and encourage their students to get actively involved in learning.

On a deeper level, GREEN participants appear to be ready to dissolve the cultural boundaries between nations to open up a greater sense of understanding, cooperation, and respect, especially as they relate to environmental issues. GREEN hopes that this will help create a more beneficial, cooperative link between cultures as the world faces serious environmental concerns today and into the future.

Cross-Cultural Partners

One of GREEN's major visions is to enable students to understand water quality issues in relation to time and space, and from an ecological, economic, political and social perspective. Students and teachers are encouraged to exchange their observations and findings with students and teachers who share similar interests about water quality issues but may be from very different backgrounds. The goal of GREEN is to provide a more meaningful learning situation for students, as well as an opportunity to form partnerships across the globe to work toward the solution of relevant environmental issues. To achieve this vision of GREEN, cross-cultural sensitivity and understanding is needed.

One important aspect of the GREEN program is the Cross-Cultural Partners Program. This program started on a pilot basis in January, 1991, with schools in Australia, Canada, Hungary, Mexico, New Zealand, Taiwan and the United States. Partnered schools shared information about their respective interests and cultures, the early settlement and development of their local watershed and rivers, visions for future watershed and river uses, and action plans and strategies needed to sustain desired uses.

Depending upon what each school wishes to emphasize in the partnership, the overall program may range from a few months to a few years. For example, if the partnership is established with students in their last year of secondary school, the students might be able to maintain the linkage for a year before they graduate and leave school. On the other hand, if students are in middle school/junior high, the partnership might continue and evolve over a number of years.

Figure 9.10. Danube River in Budapest, Hungary.

It is important that participating teachers make at least a one year time commitment. Although the actual programmatic phase might be less than a year, the teacher's involvement will be longer due to the time devoted to program planning, implementation, and evaluation.

If a partnership is established with a very specific focus in mind, the partnership might terminate at the end of one academic year. Then the teacher could repeat the whole program with a new class the next year. In such a situation, the complete partnership and cross-cultural exchange might then be planned for each group with one academic year as a time frame.

On the other hand, if a partnership can be operated as an extra-curricular activity, a teacher might be able to work with the same group of students over a number of years and thus be able to forge a long term cross-cultural linkage.

Another dimension of the program is the partnership between teachers and/or staffs of schools which become partners in educating for specific goals or programs.

The partnership-model allows for considerable flexibility and unlimited creativity and ingenuity, guided by the interests, resources, and constraints of the partners. Partners view this program as non-restrictive and open-ended, with opportunities to adjust the program to meet the needs and aspirations of those involved.

Through involvement in the GREEN network, students share information, techniques, and different approaches to problem-solving. They can learn that their investigations have a purpose and are valued by their peers elsewhere in the world. Therefore, GREEN aspires to create an international network of aware, concerned, and active students who will be prepared fully to take on the environmental challenges of today and tomorrow.

In such a network, comprised of linkages and partnerships, good communication is crucial. We are all aware of the recent rapid expansion of communication technology. However, technology is merely a tool whose existence alone will not bring about communication if there is no commitment to improve the ability to communicate. Communication will not succeed if the communicators fail to try to understand each others' perspectives and situation.

Perspectives and situations are a result of tradition, culture, and experiences. It is not enough to know that some person is engaged in some activity or that some event is occurring. It is important to know the "why" behind the action or the occurrence, and try to understand it from the perspectives of the people who are involved.

Let's look at an example. Some people in developed societies, as well as some well-to-do people in developing societies, have been known to blame rural farmers in the tropics for being the cause of massive deforestation in these countries. The first inclination of such people might be to think of those farmers as ignorant and environmentally irresponsible people who are cutting down the trees and making our planet warmer and uglier.

In reality, as is often the case, the farmers might be compelled to cut the trees because they have no other source of fuels with which to cook and keep themselves warm. Even though the farmer may recognize the environmental consequences of deforestation, lack of options might necessitate cutting down the trees today, foregoing the benefits of tomorrow. To an outsider this might seem to be an irresponsible practice that could not lead to a sustainable future, when it may in fact be a matter of life and death for the farmer's family.

Often, people tend to be judgmental and condemn another person's actions without knowing what makes that person act the way he or she does. In such instances, a cross-cultural and a contextual awareness and understanding are invaluable. The problem in the example above will not disappear or go away by pointing fingers at the desperate farmer who is cutting down a tree in order to cook a family meal or to keep the children warm. Understanding the farmer's plight, and working with him/her to resolve the problem may yield better results.

In order to achieve improved understanding and cooperation, we need to cross cultural boundaries. One of the important aspects of this program is not only to collect and send information to one's partners, but to try to understand which issues are involved, reflect on what has and is occurring, and determine the cultural biases, the consequences of actions being considered and taken, and what new approaches and actions could be successful.

With the above in mind, the overall goals and objectives of the Cross-Cultural Partners Program are:

1. *Cross-Cultural Readiness*

   Goal: To develop an interest and desire to explore and exchange information and ideas on a cross-culture level.

   Objective: To provide the opportunity, infrastructure, and training to enable students, teachers, administrators, and parents to obtain the attitudes, skills, and sensitivity to enter into a cross-cultural educational program.

2. *Cross-Cultural Background*

Goal: To develop an understanding and sensitivity to cross cultural roots, differences, and similarities.

Objective: To encourage among students the recognition of differences and similarities between members of the same, as well as different, cultural, socio-economic, or religious groups.

3. *Watershed Understanding*

Goal: To develop an understanding of past and present watershed practices and patterns on a cross-cultural, interdisciplinary level (in land-use, ecological, economic, political, social, and technological areas) that have influenced the quality and uses of particular rivers over time.

Objective: To have students collect past and present information on their local watershed that is relevant to the existing quality of their river's water.

4. *Watershed Information and Ideas Exchange*

Goal: To develop ways to collect, reflect and communicate information and ideas cross-culturally between watersheds.

Objective: To increase students' and teachers' thinking and communication skills necessary for successful cross-cultural exchange.

5. *Cross Cultural Change Processes*

Goal: To identify strategies designed to improve the water quality within each of the watersheds being studied.

Objective: To increase students' thinking about another culture's point of view, and strategies relevant to bringing about change within each of the watersheds being studied.

6. *Student Empowerment*

Goal: To empower students to take appropriate actions on a personal, school, neighborhood, community, or regional level to improve the water quality of rivers based on their cross-cultural exchange of information, ideas and aspirations.

Objective: To encourage participants to locate, assess, and identify a way to work toward the solution of at least one relevant local water issue.

7. *Working Peace System*

Goal: To work toward an arrangement to improve the quality of water in the "partners" watersheds, one that breaks away from the traditional link between authority and political boundaries, to enable people to plan together for the common good of the affected people.

Objective: To encourage diverse groups of people who are dependent upon the same international river to plan together and, in the process, to gain a greater respect for the other people involved in the partnership, a more lasting trust. Such an opening, to consider new and challenging ideas and strategies, might link people without violence to work toward a more peaceful and livable world.

8. *Cross-Cultural Evaluation Process*

Goal: To provide a comprehensive cross-cultural program evaluation which focuses on changes in areas such as: the affective, cognitive, and skill domains; critical thinking; empowerment; change strategies; and communication technology.

Objective: To provide an evaluation model that will involve students, teachers, administrators, and parents in an on-going evaluation process, and to use the information collected to improve the instructional program.

Thus, the overall goal of the cross-cultural component of the GREEN program is to enable students to develop interdisciplinary and holistic analytical skills by providing them with an interactive learning opportunity. The students learn about rivers from other cultures and societies in order to analyze their understandings of, and attitudes toward, their own rivers. Such awareness and understanding could pave the way for mutual tolerance, assistance, and cooperation, as well as directing the students in becoming more responsible global citizens.

The cross-cultural exchange component of GREEN has been developed to provide a framework (model) for sharing observations, knowledge, ideas, and solutions with people from diverse backgrounds. Typically each prospective participating school in the exchange program begins with these steps:

1. Contacting and formalizing a cross-cultural arrangement with the GREEN Project

2. Preparing students for a cross-cultural exchange

3. Developing pen-pal relationships

4. Researching the historical background and anticipated issues associated with the river system

5. Monitoring the benthic macroinvertebrates and physical-chemical quality of the river water

6. Visualizing the desired condition of the river at a time in the future

7. Identifying the laws, policies and responsibilities of regulatory agencies related to maintaining and improving river water quality

8. Developing and carrying out an appropriate action plan on an individual, school, neighborhood, community or regional level

9. Evaluating the cross-cultural sister watershed program.

The first step in becoming involved in the Cross-Cultural Partners Program starts when a school contacts the GREEN Project (see address at end of this chapter), and expresses an interest in becoming a partner in the Program. In return, GREEN sends a questionnaire to the inquirer, requesting more specific information which would be helpful in the consideration and selection of a cross-cultural partner.

The *Green Cross Cultural Partners Program Handbook* is also sent to the inquirer. The *Handbook* contains specific information regarding a framework for establishing and implementing a Cross-Cultural Partners Program, such as program goals and objectives, program design, how to get started, classroom activities, evaluation, etc.. It also contains information about the process and guidelines for selecting a cross-cultural partner (school program, class interest, existing relationship, community heritage, language, rural-urban, etc.); key elements in the Program design (developing cross-cultural sensitivity, pen-pal relationship, watershed analysis, river walk, water quality monitoring, communicating watershed/river information, identifying laws and regulating agencies and responsibilities, visualizing the future, developing an action plan, evaluation, etc.); and a formal memorandum of agreement between GREEN and the partners (tasks, timeline, funding/fund-raising, resources available, etc.)

The second step of the Cross Cultural Partners Program focuses on a series of activities for students which are designed to foster greater understanding and respect for other cultures—and in particular the culture of the emerging partnership. *The Green Cross Cultural Partners Program Handbook*

Figure 9.11. A Singapore student running a percent saturation test on the Singapore River.

contains many suggestions for ways to become more informed about another culture. Often this entails visiting a local library, interviewing particular residents in the community, viewing special video-TV-radio programs, or sending class representatives to embassies or consulates to obtain specific information on the perspective partner's society.

Also the *Cross Cultural Partners Handbook* contains activities to orient students to the cultural diversity present in the classroom; to examine cultural stereotypes; to formalize and prepare for a cross-cultural exchange; to check for bias in classroom instructional materials; and to participate in other culture building activities.

The third step involves establishing a pen-pal relationship between students in the partnered schools. Teachers ask each student to prepare a series of short letters to the participants in the partnered school (or to a particular student in the partner school). In the pilot programs students were encouraged to write about their family, neighborhood, schools, personal interests, hobbies, aspirations, life in the community, impressions of the river to be studied, etc. Each letter included a small photo of the student (clipped to the top of the letter) and often a photo of their community and/or river.

In addition to these topics, students often described other aspects of their personal lives. In the classroom, teachers and students discussed how they would describe their community to someone who had never visited their region or nation. It was emphasized that pen-pal letters would be sent from each class to their partners at approximately the same time. If translation was necessary, then the schools used the skills of students in their language classes or community residents to translate information.

The fourth step of the Cross-Cultural Partners Program involved students in becoming much more informed as to past and present watershed practices. Developing this understanding usually involved five steps. It was suggested that the class divide into sub-groups and that each group take responsibility for one component.

The activity involved researching and organizing watershed information to be sent to the sister school by mail, computer, short-wave radio, FAX or traveler. The areas of focus were:

1. Gathering information about the physical characteristics of the watershed and river including its length, size, location of population centers, soils, topography, special features, the location of the school on the watershed, and other pertinent information;

2. Reviewing the early human settlement patterns within the watershed and along the river, its use by native peoples and early settlers, and the origin of the river's name;

3. Researching the past and current uses of the watershed and river; for example, agriculture, logging, fishing, urbanization patterns, municipal and industrial uses, and recreational opportunities.

4. Identifying and researching current patterns that have implications for changes in water quality, and critical issues associated with the watershed and river water quality, including those that are concerns of the students.

Figure 9.12. A Hungarian student helping to evaluate the international aspects of the water monitoring program.

5. Describing people's hopes and visions for their local river in the future, including the aspirations of the students involved in the program.

Students gathered the information using newspaper articles, historic records, journals, focused interviews, recordings, tapes, museum exhibits and material, and record of societies and various community and state organizations and institutions. The students had an opportunity to gather information from local and elderly people who know the history of the river through personal experiences. The information each group developed was edited and sent to the sister watershed group through appropriate communication channels.

The fifth step encouraged students to walk along the river to record land-use practices, and monitor the river for benthic macroinvertebrates and physical-chemical parameters. (Both partner school groups performed the same land-use activities and water quality tests). The information was communicated to partner schools by the most appropriate communication channels. The classes then spent considerable time gathering information from other up- and downstream locations, frequently graphing the material by hand or computer, and reflecting on what the results indicated.

The sixth step consisted of the students obtaining information on the laws, policies, standards and responsibilities of water quality regulating agencies on a federal, regional, state, county, or local level. Usually this material was formalized into a chart and available for class members interested in contacting organizations and decision-makers when carrying out their action plans. This information was communicated to their cross-cultural partners.

The seventh step focused on visualization—a technique used by people to help solve problems and to envision solutions. In reference to a local river or stream, students developed a vision of the watercourse they were studying. Students focused on what their hopes and aspirations are for the uses, conditions, and plans for their river in the future. After their prior experiences with their watershed through interviewing, monitoring, and considering trends and future opportunities, the students generated more comprehensive ideas regarding uses and the future of their river.

The eighth step addressed the students' action plan on individual, school, neighborhood, community, and regional level. Rivers face many problems, and these problems often cause feelings of helplessness and apathy among people. It was, therefore, important that students be given opportunities to become involved in meaningful actions to create positive changes.

The previous activities in this section extended students' knowledge about rivers, awakened their concerns, provided tools for gathering more information, and generated ideas for change. This action plan was designed to be a culmination of those activities; using skills developed, it was an opportunity to take some appropriate action at a level that was relevant to the students.

The ninth, and final step, was a student-teacher-administrator-parent evaluation of the total program and suggestions for improvement. Evaluation provided important feedback to students, teachers, administrators, and parents. The results helped to determine how effective the learning experience was, to identify the strengths and weaknesses of the program, and to provide suggestions for improvement.

Evaluation is critical in all educational endeavors. There are two aspects of the partners watershed program being evaluated. The first is the function of the partnership and the strength of the linkage. The second focus of the evaluation is to assess changes in the student, in teaching practices, and in the institution as a result of the program. The latter evaluation could focus as well on changes in the affective, cognitive, and skill domains of students; empowerment of students/teachers; internal and external institutional changes; etc.

Each stage in the Cross Cultural Partners Program lasted approximately one month. The time line was offered as a guide to keep the exchanges in a similar time frame. However, it was made clear that the time line was only a suggestion and the partners should adjust it to meet their educational needs.

Cross-Cultural Partners in Action

During the first year of exchange, the following were some of the noteworthy events that occurred:

Barcs High School (Barcs, Hungary)

Students collected photos, articles, and historic documents on two local rivers—the Drava and Rinya; obtained detailed topographic and hydrometric watershed data; produced homemade videos on the rivers; and monitored for benthic macroinvertebrates and physical-chemical parameters. Their works and findings will be used to produce a public exhibit to broaden public awareness and understanding of the importance of these rivers, and of the problems that need to be addressed.

Cleveland School of Science (Cleveland, USA)

Students contacted and interviewed Hungarian residents in the Cleveland area regarding the immigration of their ancestors to the Cleveland region during the industrial revolution to work in the steel mills. Hungarian residents also volunteered their services to assist in the translation. The Cleveland School of Science hosted the Cuyahoga River Student Congress and shared with participants their experiences with the Cross Cultural Partners Program with Barcs High School.

Loungsan Junior High School (Taipei, Taiwan)

A formal slide and written presentation was prepared and delivered by students and teachers to the Taiwan Environmental Protection Agency, including results of their investigations and a set of recommendations to improve water quality on the Tanshuei River. A formal report on the product is also being prepared for submission to the Taiwan National Science Council, one of the sponsoring organizations of the Cross-Cultural Partners Program.

I **Figure 9.13.** Using the river for recreation.

Model High School (Bloomfield Hills, Michigan, USA)

This school focused their water quality monitoring program on two sections of the river, upstream and downstream from a new construction site. Students and the teacher prepared a written transcript and made a slide presentation to the Bloomfield Hills Board of Education about this project and their partnership with a sister watershed in Taiwan. As a result, the Board became more knowledgeable about the dimension and potential of this educational program, and provided support to continue the inter-disciplinary Cross Cultural Partners Program as a model for cross-disciplinary teaching in the school district.

Sydney High Schools (Sydney, Australia)

Streamwatch is a growing program that will eventually involve students from up to 200 high schools throughout the Sydney metropolitan region. Monitoring information is communicated periodically to Streamwatch for dissemination and directly to the Sydney Water Board for follow-up action. Streamwatch published a *Special Environmental Education Newsletter* providing timely information on river water quality issues. The Sydney Water Board recently appropriated $850,000 to equip Streamwatch schools to monitor four major catchment areas in the Sydney metropolitan area.

Saginaw High School (Saginaw, Michigan, USA)

Throughout the school year, a student "River Cast Team" produced a monthly 15–20 minute video about their Interdisciplinary Saginaw River Interactive Water Quality Program. A media/ broadcasting class provided regular five-minute radio interviews throughout the year focusing on history, events and issues related to the Saginaw River. The student "River Journal" team prepared and published in their community newspaper a 24–page *Saginaw River Journal*, which included student-written articles about history, water quality test results, and student actions.

Following the Saginaw River Student Congress, student representatives presented to the City Council their annual water quality data on the Saginaw River. The fecal coliform data in particular indicated conditions unsafe for the Annual Raft Race down the Saginaw River. The Saginaw City Council responded by approving a new set of safety rules and designating another section of the Saginaw River for the event.

Five elementary school classes authored and published documents (40–50 pages each) about the river, including "Rivers of the World" and "Saginaw River Poetry." Students prepared an article entitled "Water Conservation Tips," for their community paper. The Arthur Hill High School hosted the Saginaw-Flint River Student Congress and provided a workshop on the Cross-cultural Partners watershed program.

New Zealand High Schools (New Zealand)

The New Zealand Natural Heritage Foundation has been very effective in developing a network of schools to monitor the Manawatu, Oroua, Wanganui, and Ngaruroro Rivers. The product has received attention from national and community newspapers, benefited from EARTHWATCH radio programs, received local watershed council support, and is in the process of establishing a computer network of watershed coordinators throughout New Zealand on EcoNet.

The New Zealand Natural History Foundation also hosted an international environmental education conference in August, 1991 with a major focus on cross-cultural watershed exchanges. The New Zealand Cross-Cultural Partners Program received funding from four sources and two prestigious awards during its initial year.

Bellevue High School (Bellevue, Michigan, USA)

Students requested, and received permission, that the City lower the water level to allow students to collect polluting materials and other debris along the shore. Students also sponsored a successful Community Household Hazardous Waste Cleanup Campaign.

Rio Grande High Schools (Mexico and U.S.)

For much of its 1800–mile length, the Rio Grande forms the international boundary between the U.S. And Mexico. It is appropriate, then, that high schools in both the U.S. And Mexico participate in Project del Rio, a student water monitoring program in the Rio Grande watershed. Students use a bilingual computer network to communicate with other schools. In its pilot year in 1991, Project del Rio was the topic of a national television news broadcast and received considerable local media coverage. The project is expanding, and will eventually include schools along the entire length of the Rio Grande. Through student water monitoring, Project del Rio is helping to foster cross-cultural understanding and cooperation among U.S. and Mexican citizens as they work to protect their common environment.

Dearborn Edsel Ford High School (Dearborn, Michigan, USA)

Dearborn Edsel Ford High School hosted the Rouge River Student Congress and provided a special workshop for students and teachers on their Cross Cultural Partners Program with New Zealand.

Kawai Nui Marsh: Hawaiian Case Study

The Global Rivers Environmental Education Network (GREEN) was invited to Hawaii to work with Aikahi Elementary School in Kailua to help focus their integrated curriculum around the famous and sacred Kawai Nui Marsh.

The Kawai Nui Marsh is situated in the Maunawili Valley on the windward side of the island of Oahu. The geologic features of this valley were primarily formed by the lava flows of the Koolau and Kailua volcanic series. At the base of the high cliffs of the Koolau Range, deep in the valley, older alluvium (material deposited by running water in the past) forms an apron, while the lower part of the valley is underlain by younger alluvium.

This younger material forms an important geographic feature in the valley: Kawai Nui Marsh, the largest freshwater marsh in Hawaii. It has been suggested that Kawai Nui Marsh was once a freshwater lake transformed by siltation into the marsh that exists today. Along the front of the marsh, as along the entire coastline, there are sand dunes and outcroppings of limestone.

The marsh is presently designated as a flood control conservation area and habitat for four endangered species: the Hawaiian duck; Hawaiian stilt; Hawaiian gallinule; and the Hawaiian coot. In addition, the Kawai Nui Marsh serves to recharge the groundwater, acts to retain sediments and nutrients, provides habitat for fish, and is used for recreation. Ancient Hawaiian culture has left a legacy of understanding and sensitivity for nature.

Figure 9.14. Kawai Nui Marsh

A dike separates the Kawai Nui Marsh from the community of Coconut Grove, which suffered a serious flood in 1988 when 23 inches of rain fell in the upper watershed within a 24-hour period. The marsh drains into Kailua Bay, one of the world's finest beach areas and location of some of the most biologically diverse coral reefs in the world.

Three streams drain into Kawai Nui Marsh: the Maunawili, the Kahanaiki, and the Ka'pa. It is in this area where two world class golf courses are being built and two subdivisions. The water runoff from these developments poses a threat to the water quality and diversity of plants and animals found in Kawai Nui Marsh and Kailua Bay. It also presents a flood threat to Coconut Grove.

During the ten days that GREEN worked with Aikahi Elementary School, students from all grades were oriented to the cultural history, geology, ecology and recent human developments in the area. Students, teachers and parents were bussed to the crest of the Maunawili Valley, and each ecosystem in the watershed was studied, including: tropical rainforest, river systems, subdivisions, golf courses, cultural sites, Kawai Nui Marsh, the flood dike, the Coconut Grove community, two canal systems, and Kailua Bay.

The students recorded impressions of each ecosystem on their art pads and on video, identified the benthic organisms in the connecting river system, analyzed water samples, studied historical cultural sites, and scrutinized the existing laws and regulations that govern the area.

Figure 9.15. An elementary student drawing a portion of the watershed on her art pad.

The administrators, teachers, and students of Aikahi Elementary school analyzed their watershed information and presented their findings and recommendations to city officials, environmental groups, area residents, school officials, and parents.

The elementary, middle, and secondary schools of Kailua have joined together to establish a comprehensive water quality curriculum, involving the entire watershed within their district. The curriculum will include: cultural history, benthic macroinvertebrates, water quality testing (dissolved oxygen, fecal coliform, pH, biochemical oxygen demand, temperature, nitrates, total phosphates, turbidity, and total solids), changes in land use practices, existing laws and regulations governing the watershed, computer

Figure 9.16. A teacher recording the benthic macroin-vertebrates identified by students on her portable computer.

conferencing between schools and regulatory agencies, and the taking of appropriate actions to prevent further damage to the Kawai Nui Marsh ecosystem.

The teachers, students and administrators have been asked to make presentations to schools and communities throughout the Hawaiian islands. Their efforts are starting to spawn other interdisciplinary school programs, hopefully resulting in more ecologically sound planning in other island communities.*

*References: Kawai Nui Heritage Foundation, P.O. Box 1101, Kailua Hawaii 96734; and State of Hawaii, Department of Land and Natural Resources, Honolulu Hawaii.

Walpole Island: Native American Case Study

The Walpole Island First Nation Reserve is in southeastern Ontario, located on a freshwater delta at the mouth of the St. Clair River. This delta is an accumulation of sand and silt deposited over several thousand years. The islands have been built up on the clay plains which form the basin of Lake St. Clair in the Great Lakes ecosystem.

Totaling 58,000 acres, Walpole Island contains a rich mosaic of ecosystems: tallgrass prairies, oak savannas, and a diverse system of wetlands. Approximately one-third of the reserve is dryland, containing agriculture, forest, savanna, and prairie. Two-thirds is wetland, including marshes, sloughs, interior lakes, channels, and adjacent Lake St. Clair waters.

Figure 9.17. Walpole Island is the large marsh area on the large delta extending into Lake St. Clair on the eastside of the St. Clair River.

Cultural History

Walpole Island has been inhabited by native peoples for more than 10,000 years. The three tribes currently residing on Walpole Island are of Algonquian language descent, and their history in this area is extensive. At one time, Ottawas, Ojibwa, and Potawatomis were all living near the Straits of Mackinaw. In the mid-fifteenth century, the Potawatomi and part of the Ottawa migrated south, where they relied primarily on agriculture. The northern Ojibwa and Ottawa bands were fishers, hunters and gatherers, migrating seasonally.

Today, Walpole Island First Nation is governed by an elected Chief and 12 Councilors. Its population of approximately 2,200 Ojibwa, Ottawa, and Potawatomi people has united a political and cultural confederacy called the Council of Three Fires. Although Walpole Island First Nation has had much contact with surrounding communities, its members have retained an Indian identity at the level of their deepest values and ways of living.

Relationship To The Environment

Through the years the bands have practiced careful stewardship of the land. Their management has not resulted in depletion of natural resources or environmental degradation. To Walpole Island First Nation, "the land" includes not only dry land, but air, water, plants, animals,

Figure 9.18. Part of the marsh environment of Walpole Island, that appears to be relatively undisturbed following 10,000 years of habitation.

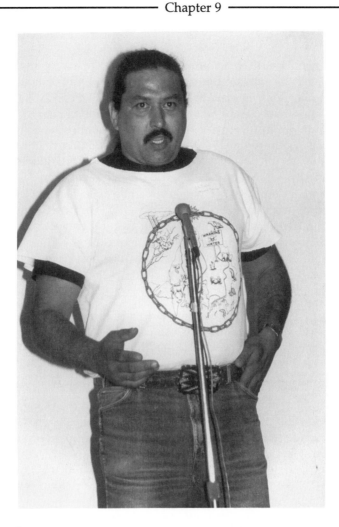

Figure 9.19. Mr. Ed Issacs of the Walpole Island Community informing students at the Black River Congress about the concerns of his people in the water quality of the Great Lakes.

humans, and a corresponding responsibility to protect these elements. For Ojimwa, Ottawa, and Potawatomi, the land has social, cultural, and spiritual significance, as well as economic importance. The people of Walpole have a very close relationship with the land, and they have traditionally relied to a significant extent on local resources. Consequently, the health of the Walpole community and culture is closely tied to the health of the local environment.

Sustainable Development

To the Walpole Island First Nation, sustainable development implies social, cultural, spiritual, environmental, and economic well-being. The band has defined sustainable development as "the process of equitable economic, social, cultural, and technological betterment in a way that does not pollute ecosystems and irrevocably deplete resources." A guiding concept for Walpole Island First Nation is that the land and its resources are to be preserved for the benefit of past, present, and future generations. Sustainability has continued to be a way of life for residents.

Program Background

In May 1992 a student in the Black River Water Quality Monitoring Program went to Walpole Island to personally invite the Walpole community to attend the Black River Student Congress.

During the student congress members from Walpole Island participated fully in student and community presentations, took extra time to learn how to monitor for water quality, and participated in computer training sessions.

Walpole representatives provided the opportunity for students at the Black River Congress to attend the Annual Indian Ceremonial Celebration on Walpole Island. Several students participated in the ceremony and learned more about the culture of Walpole First Nation People.

Figure 9.20. Walpole Island students learning how to use the water quality tests during a workshop session.

In the fall of 1992, GREEN contacted Dean Jacobs, Director of the Walpole Island Heritage Centre and member of the Ontario Round Table on Environment and Economy, to determine interest in involving the Walpole Island School in a water quality educational program. The mission of the Heritage Centre is to preserve, interpret, and promote the natural and cultural heritage of the Walpole Island First Nation Community. Nin Da Waab Jig, meaning "those who seek to find," is the phase which captures the essence of the Heritage Centre's work.

Figure 9.21. Testing the water that flows through Walpole Island.

GREEN was asked to provide a presentation on the purpose and structure of a water quality monitoring program to the Chief, members of the Council, elders, residents, students and teachers, and members of the Walpole Island Heritage Centre. Following this workshop, the staff of the Walpole Island Heritage Centre informed GREEN that they would be interested in involving their community in a schoolwide water quality monitoring program.

Program Design

The water quality monitoring program for the school community of Walpole Island was designed around content and structure similar to other GREEN programs throughout North America and the global community. The program was developed in cooperation with the Walpole First Nation People. It involved watershed concepts, historical uses of land, land management, water quality monitoring and analysis, water laws and regulations, interactive telecommunication, and a student-community congress. The program also included instructional materials designed to enrich the existing school curriculum, with the full involvement of the teaching staff of Walpole Island Elementary intermediate school and Wallaceburg High School.

In its second year, the program expanded the computer telecommunication linkage to include indigenous and non-indigenous schools throughout the Great Lakes region and elsewhere in Canada and the United States. Responding to local concerns and to the water quality data gathered in the first year, the program also began to focus on heavy metal monitoring.

Na'aman River: Arab and Jewish Israeli Case Study

GREEN and the Nature Reserve Authority of Israel embarked on a new project in November, 1993. This program was designed to bring together Jewish and Arab Israeli students within the same watershed to work cooperatively to improve the quality and quantity of their water and their cultural relationship.

The first phase of the program established a common water monitoring program, involving 15 secondary school classes from both Arab and Jewish schools in the watershed of the Na'aman River, north of Haifa, Israel. The second phase was designed to address critical water issues on the Na'aman River.

The Nature Reserve Authority (NRA) is a governmental organization which manages the nature reserve system of Israel (approximately 400 reserves with 50,000 hectares throughout the country). NRA is responsible for the preservation of nature and for the enforcement of hunting and environmental laws in open areas (in conjunction with other governmental

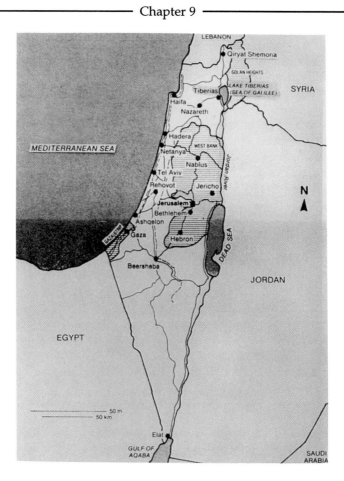

Figure 9.22. The Na'aman River is located north of Haifa, Israel.

authorities). NRA is also participating in the development of educational resources for nature conservation and landscape protection, in accordance with the Ministry of Education's curriculum requirements.

Seven guiding centers located in specific reserves are currently in operation, each one responsible for a specific region, such as the En Afeq Guiding Center in western Galilee. The En Afeq Center provides field activities for classes of all ages, including Muslim, Druse, Greek Orthodox and Jewish students from local schools. The Center is located in an 800-year old Crusader watermill, refurbished by the NRA.

In addition to its current activities, the En Afeq GREEN Program provides the opportunity for students from both Arab and Jewish high schools to monitor the waters of the Na'aman River, resulting in the establishment of database information and watershed management and protection strategies.

Environmental protection and wise use of natural resources is critical in every comer of the world. In the Middle East, the scarcity of water resources, increased pollution, and depletion of available water are all leading to a grave water crisis that must be addressed by all citizens, regardless of ethnic, religious, or cultural affiliation. Citizen efforts to conserve precious water resources at the local level is a critical element of any water management strategy.

Educating students about water and involving them in solving real problems is an excellent step toward achieving this goal. In addition, a program that brings students from diverse cultural and religious groups together to address mutual problems provides a foundation for continued future cooperation. Efforts to bring Jewish and Arab Israeli students together for cooperative programs have been successful in the past, but have not been done in an environmental context. This program combines these two important educational goals in a real-life, immediately useful activity.

Program Goals

➤ Students become empowered to help protect their local water resources by introducing them to water monitoring in a hands-on, action-oriented program.

➤ Students learn core scientific skills and concepts, such as ecological cycles, food chains, water chemistry, aquatic biology, and the interaction of land use and water quality. They also learn social science such as the geography of the watershed, the history of land and water use in the area, the economics of the local use of natural resources, and the government processes involved in managing natural resources, including an understanding the laws, regulations, and regulatory agencies.

➤ Students develop interpersonal, communication, and problem-solving skills by working in teams to test the water, interpret the results, identify problems, and search for solutions. This cooperative effort is enhanced by working in English, the second language common to both groups.

➤ Arab and Jewish Israeli students have a forum to work cooperatively on relevant problems of mutual concern.

In addition, students provide a service to the community by addressing important issues concerning water quality and quantity.

Program Content

The program includes well-developed and tested instructional materials consisting of watershed maps, an instruction manual for conducting water quality tests safely and accurately, an instructional material for sampling benthic macroinvertebrates as indicators of water quality, and water quality testing kits for nine water quality tests.

Students learn about their watershed, learn to conduct the water quality tests, do the field work, analyze their data, calculate a water quality index, and analyze the ecological and human factors that result in the overall water quality they obtained. This analysis includes exploration of the watershed and land uses by examining satellite images dating back to 1972 to determine the impact of changing land-use practices on the river system. Then the students address critical water quality and quantity issues of mutual concern by helping devise management plans or action projects. Throughout the program, students from all participating schools—Jewish and Arab—work together to achieve common goals.

Thirty classes from five schools (three Jewish and two Arab) will participate in the GREEN program at En Afeq in the first year. The program is directed by Dr. Reuven Ortal, with onsite supervision and implementation by Ra'aia Shoorky (reserve manager) and Mahmud Nassar. The actual teaching is done by a team of part-time guides consisting of ecology and biology students (3rd-year or higher) at the University of Haifa.

Figure 9.23. Arab-Israeli students being oriented to the Na'aman River.

The guides are both Arab and Jewish; the Arab guides are fluent in Hebrew as well as Arabic. Jewish classes are taught by Jewish guides and the program is conducted in Hebrew; Arab classes are taught by Arab guides and the program is conducted in Arabic. This is the first educational program in Israel in which activities at a nature reserve are being provided to Arab students in their first language, demonstrating to both the students and teachers that their participation and contribution is valued and desired.

The program consists of three parts: classroom work, field work, and a final meeting. The program is designed to fulfill the requirements of the Ministry of Education's ecology program for the high school level. This program requires 10 hours of classroom work in ecology plus three field trips. Accordingly, each class comes to the reserve for three field days lasting four or five hours each. These are scheduled one each for fall, winter and spring, enabling students to observe seasonal changes. In addition, students receive five classrooms sessions of two hours each, some of which take place just prior to the field trips in order to prepare for the work in the field.

On the first field visit, which takes place in the fall, students tour the reserve in small groups, recording their observations of flora and fauna, weather, appearance of the water, etc. This tour focuses on the clear springs and ponds that form the headwaters of the Na'aman River.

Figure 9.24. Jewish-Israeli students studying land use practices along the Na'aman River.

Each group stops at a different site along the springs and conducts chemical and physical tests, including width, depth, and velocity; temperature, dissolved oxygen, pH, conductivity (a measure of salinity), clarity, ammonia, nitrates, and phosphates. These tests were chosen based on the parameters used by the Water Pollution Control Unit of the Nature Reserves Authority in the enforcement of water quality standards in Israel. This slight variation from the nine tests that make up the overall WQI used in this manual makes the program consistent with Israel's governmental water monitoring, rather than using the parameters chosen by the U.S. This is a good example of the flexibility of the GREEN program in practice.

When the testing is complete, the groups reconvene at the guiding center and travel by bus to a site outside the reserve. This site is along the Na'aman River in an industrial area. A strong, unpleasant odor permeates the area, and the most dominant feature is a stagnant sewage pond with an unnatural red color and a spillway to the river. Again, the students record site information and draw the site individually, and test the water for the physical and chemical parameters.

The second visit takes place during the winter, about a month and a half following the first visit. About six weeks prior to this visit, benthic macroinvertebrate and plant samplers are placed in the water at both sites—enough to have one for all 30 classes at both sites. Just prior to the visit, the benthic sampler for the class is removed and samples are prepared in petri dishes in the lab in the old flour mill.

Figure 9.25. Staff of the Na'aman River Program draw from the Druse, Christians, Muslim, Jewish and Bedouin communities in Israel.

During this visit, the students again tour the reserve and observe seasonal changes. They also conduct a brief survey of land animals and birds they observe in the area, measure the rate of decay in the water of different plant species from the plant samplers, and collect and briefly observe benthic organisms. Then they work with stereoscopic microscopes in the lab to obtain a sequential comparison index of benthic organisms from the prepared samples, using a background grid in the petri dish to ensure a random sample.

The third visit, devoted to data analysis with the help of computers, takes place in the spring. Again, students tour the reserve and record observations, noting seasonal changes. Then, in the computer lab at the renovated 12th-century Crusader flour mill, they enter their data into the computer database created for the program. With each class contributing data from its particular days at the reserve, data for the chemical tests will be available for the entire fall, and data for the benthic macroinvertebrate index will be available for the entire winter. The cumulative results of the students' efforts is a complete set of data collected on an on-going basis.

During the third visit, students analyze this data and draw conclusions about the quality of the Na'aman River at the two sites. This analysis can include data tables and/or graphs comparing the two sites on the particular days their class visited the reserve, comparison of the two sites over the entire seasons, and in the future, comparison of the two sites over the entire seasons and from year to year. In addition, students may have access to data from the past 20 years, collected by Nature Reserve Authority biologists. This will enable them to compare their results with trends from the past to see if they detect any unusual changes. If so, they could then suggest possible reasons for these changes, such as accuracy of their testing or real environmental changes, either human or natural.

Following this data analysis, students will research some problem affecting water quality on the Na'aman river, based on their observations and water monitoring results. These problems normally fall within three general categories—industrial sewage, agriculture discharges, and domestic sewage—but within these categories, students may choose any problem on which they wish to focus.

For example, within the category of agriculture, they may focus on drainage of cotton fields or aquaculture ponds at the nearby kibbutz. They will interview agency officials, industry managers, village council members, kibbutz managers, etc. to learn about details of the activities contributing to the problem, including economic, social and cultural constraints that must be considered in effecting change in these areas.

The project will conclude with a final report from each student about what they have discovered about the water quality of the Na'aman river, both in the reserve and downstream from it, and the problem they researched.

In addition, the joint cooperation element of the program will be implemented by the teachers in the final part of the program, and will vary from class to class depending on the decisions of the teachers. Two or more Jewish and Arab classes may get together to work on a problem to be jointly researched, or they may come together for the final report stage to share information and learn from each others' work. This aspect of the program promises to evolve over the next few years as teachers gain experience with the program and the Arab and Jewish teachers develop good working relationships with each other.

If you are interested in obtaining further information about GREEN, please contact:

Earth Force
1908 Mt. Vernon Ave.
Alexandria, VA 22301

Telephone: 703-299-9400
Fax: 703-299-9485
World Wide Web: www.earthforce.org
Email: green@earthforce.org

GREEN and the World Wide Web

The World Wide Web offers students a way to share water quality information, research water quality issues, and communicate with others via the Internet. GREEN has a Web site: http://www.earthforce.org. The GREEN Web site provides information about GREEN programs around the world. Also, users have easy access through links to watershed education programs globally.

Another way to get connected with other organizations and persons engaged in environmental education is through EE-Link at http://eelink.umich.edu. This is an environmental education Web site developed through the National Consortium for Environmental Education and funded by the U.S. Environmental Protection Agency. The site houses information about projects, materials, and information about environmental education.

Appendix A
Water Quality Testing Equipment

The majority of supplies that you need to conduct the tests in this manual are available through GREEN. To order contact

Earth Force
1908 Mount Vernon Ave.
2nd Floor
Alexandria, VA 22301
Tel: 703-299-9400
FAX: 703-299-9485
Email: green@earthforce.org
Internet: <http://www.earthforce.org>

The following companies carry a number of water quality testing kits and each has a catalog that can be ordered for more specific information. This is not an exhaustive list.

LaMotte
LaMotte Company
P.O. Box 329
802 Washington Avenue
Chestertown, MD 21620
Tel: 1-800-344-3100, 410-778-3100
FAX: 410-778-6394
E-mail: ese@lamotte.com
Internet: http://www.lamotte.com

Millipore
80 Ashby Road
Bedford, MA 01730
Tel: 1-800-221-1975, 617-275-9200
FAX: 617-275-5550

Ohmicron
375 Pheasant Run
Newtown Industrial Commons
Newtown, PA 18940
Tel: 1-800-544-8881, 215-860-5115
Fax: 215-860-5213

Water Quality Testing Equipment: Range, Accuracy, Price, and Ordering Information

Range, Accuracy and Price

MANUFACTURER	INSTRUMENT	RANGE	ACCURACY	PRICE
pH				
LaMotte	pH Kit (2119)	5.0–8.0 pH units	0.5 units	$28.00 (US)
Dissolved Oxygen				
LaMotte	Test Kit (5860)	NA	0.2 mg/L	$35.70 (US)
LaMotte	Test Kit (7414)	NA	0.2 mg/L	$35.70 (US)
Nitrogen				
LaMotte	Nitrate (3354)	0.0–15.0 mg/L		
LaMotte	Test Kit (3110) Cd	0.25–10 ppm		$56.00 (US)
LaMotte	Test Kit (3354) Zn	0.0–15.0 ppm		$38.60 (US)
LaMotte	Test Kit (3615) Cd	0.0–1.0 ppm		$79.80 (US)
Ohmicron	Nitrate Screen	0.25–25 mg/L (color chart)		$99.00 (US)
		0.06–5.75 mg/L (spectrophotometric)		
Phosphorus				
LaMotte	Test Kit (7884 & 3121)	0.0–2.0 ppm		$153.35 (US)
Turbidity				
LaMotte	Test Kit (7519)	5–200 JTU	5 JTU	$37.20 (US)
Temperature				
LaMotte	Thermometer (1066)	-5–45°C		$17.50 (US)

Fecal Coliform Supplies

MILLIPORE

Filter Holder/Sterifil System	XX1104700	$62.00 (US)
Filters/type HC	HCWG0473	$80.00 (US)
Petri dishes w/pads	PD10047SO	$24.00 (US)
Syringe/hand vacuum pump	XKEM00107	$35.00 (US)
Fecal coliform media Pkg. of 50	M00000P2F	
Total coliform media	M0000002E	$20.00 (US)

Appendix B
Handouts, Data Sheets and Surveys

CATCHMENT ASSESSMENT COVER SHEET

1. Evaluator's Name(s): _____

2. Date:_____

3. Country: _____

4. State, Region or Province: _____

5. Nearest City, Town or Village: _____

6. Catchment/Watershed:_____

7. Name of Stream or River (or other water body): _____

8. Observation Points: _____

9. Survey Reach Length (m): _____

10. Reason for Survey and Goals of Catchment Assessment Activities:

_____

_____

_____

_____

_____

_____

_____

_____

_____

_____

_____

_____

_____

_____

DATA SHEET 1
MAPPING A CATCHMENT

Watercourse/Water Bodies

1. Area of Catchment: _____

2. Number of branches within the catchment: _____

3. Watercourse name(s): _____

4. Types of watercourses:

   ☐ ephemeral   ☐ intermittent   ☐ perennial

5. Average watercourse gradientv _____m/km

6. Discharge: _____m³/sec

Local Land Use in Catchment

7. Land uses in catchment (check all that apply in appropriate categories and add percentages if known or use best estimate of percentage):

   Agriculture:

   ☐ Pasture/grazing   ☐ Row Crops   ☐ Orchards

   ☐ Dry-Cropping   ☐ Agroforestry   ☐ Pesticide/Herbicide Use

   ☐ No Till   ☐ Till   ☐ Grains (wheat, oats, etc)

   ☐ Other (explain) _____

   Total Agricultural Area Percentage _____

   Urban/Suburban areas:

   ☐ Residential   ☐ Commercial   ☐ Golf Courses

   ☐ Other (explain) _____

   Total Urban/Suburban Area Percentage _____

Industrial areas (specify):

_____

_____

Total Industrial Area Percentage _____

Mining:

☐ Surface   ☐ Deep

☐ Other (explain) _____

Total Mining Area Percentage: _____

Logging:

☐ Clearcut   ☐ Selective cut

☐ Other (explain) _____

Total Logging Area Percentage: _____

Grassland:

☐ Grassland

Total Grassland Area Percentage: _____

Forested land:

☐ Forested land

Total Forested Land Area Percentage: _____

Other uses (explain; for example, sanitary landfill):

_____

_____

_____

_____

_____

_____

_____

Total Area Not Elsewhere Classified: _____

(NOTE: Percentages must add up to 100)

DATA SHEET 2

BANK AND RIPARIAN VEGETATION EVALUATION

Describe below the natural physical surroundings of the catchment you are observing.

Bank vegetation:

☐ Barren   ☐ Grasses   ☐ Brush

☐ Deciduous   ☐ Conifer   ☐ Other

Species of bank vegetation:

Herbaceous	Woody
_____	_____
_____	_____
_____	_____
_____	_____
_____	_____
_____	_____
_____	_____
_____	_____

Use the space below for drawing:

BANK VEGETATION   **SCORE** _____

4 (excellent)	Native vegetation in undisturbed state
3 (good)	Mostly native vegetation mildly disturbed
2 (fair)	Native vegetation moderately disturbed
1 (poor)	Exotics, native vegetation severely disturbed

Riparian vegetation:

☐ Barren ☐ Grasses ☐ Brush ☐ Deciduous ☐ Conifer

☐ Other_____

Species of riparian vegetation:

Herbaceous	Woody
_____	_____
_____	_____
_____	_____
_____	_____
_____	_____
_____	_____
_____	_____

Use the space below for drawing:

RIPARIAN VEGETATION **SCORE** _____

4 (excellent)	Native vegetation present/canopy intact
3 (good)	Mostly native vegetation/canopy virtually intact
2 (fair)	Native vegetation clearly disturbed
1 (poor)	Exotics/cleared land or urban development

DATA SHEET 3

BANK EROSION/STABILITY EVALUATION

1. Estimate the percentage of bare soil: _____%

2. Bank slope:

    ☐ Steep   ☐ Moderate   ☐ Slight

3. Bank stability:

    ☐ Stable   ☐ Slightly eroded   ☐ Moderately eroded

    ☐ Severely eroded

EXTENT OF SLUMPING AND MOVEMENT   SCORE _____
  4 (excellent)   No movement
  3 (good)   Slight movement on the banks
  2 (fair)   Moderate bank collapses
  1 (poor)   Severe bank failure with extensive cracking and fall-ins

AMOUNT OF BANK EROSION   SCORE _____
  4 (excellent)   Stable, no sign of any bank erosion
  3 (good)   Very occasional and very local erosion
  2 (fair)   Some erosion evident
  1 (poor)   Extensive erosion

DATA SHEET 4

PHYSICAL CHARACTERISTICS

1. Location Point: _____

2. Stream Type:

   ☐ Straight   ☐ Meandering   ☐ Braided

   ☐ Channelized   ☐ Pool/Riffle

3. Today's Weather: _____

4. Last Precipitation:

   Date _____

   Amount (mm) _____

   Duration (hours) _____

5. Recent Weather: _____

   (For data requiring more than one sample, each sample should be taken by different students/participants and the samples should then be averaged for the final figure.)

	Average	Sample 1	Sample 2	Sample 3	Sample 4	Sample 5

6. Air Temp: _____   _____   _____   _____   _____   _____

7. Water Temp: _____   _____   _____   _____   _____   _____

Physical Characteristics

	Average	Sample 1	Sample 2	Sample 3	Sample 4	Sample 5

8. Stream width (m):

   _____   _____   _____   _____   _____   _____

9. Stream depth (m):

   _____   _____   _____   _____   _____   _____

10. Surface velocity (m/sec):

   _____   _____   _____   _____   _____   _____

	Average	Sample 1	Sample 2	Sample 3	Sample 4	Sample 5

11. Bankfull width (m):

(a) ___ ___ ___ ___ ___ ___

(b) ___ ___ ___ ___ ___ ___

(c) ___ ___ ___ ___ ___ ___

(bw) ___ ___ ___ ___ ___ ___

12. Channel slope (m/km):

___ ___ ___ ___ ___ ___

13. Channel cross-section:

☐ rectangular ☐ U-shaped ☐ V-shaped ☐ other _____

14. Watercourse bottom, predominant type:

Inorganic:		*Organic:*
☐ bedrock	☐ boulder	☐ muck-mud (black, very fine organic)
☐ cobble	☐ gravel	☐ pulpy peat (unrecognizable plant parts)
☐ sand	☐ silt	☐ fibrous peat (partially decomposed plants)
☐ clay		☐ detritus (sticks, wood, coarse plant material)
☐ other _____		☐ logs, limbs

Substrate composition:

% inorganic _____   % organic _____

15. Frequency of flooding (if known or best estimate):

☐ none   ☐ rare (10 to 20 years)

☐ occasional (5 to 10 years)   ☐ frequent (1 to 5 years)

☐ seasonal

16. Watercourse channel alteration including dates (if known):

☐ dredged _____   ☐ channelized _____

☐ dam/weir _____   ☐ wetland drainage ____

☐ other _____

DATA SHEET 5

PRIMARY USES AND IMPAIRMENTS

1. Population served: _____

2. Catchment primary uses by humans:

   ☐ Domestic drinking water supply

   ☐ Bathing

   ☐ Recreation:

       ☐ Swimming   ☐ Fishing

       ☐ Other (explain) _____

   ☐ Washing clothes

   ☐ Agricultural water supply:

       ☐ Irrigation   ☐ Livestock   ☐ Other (explain) _____

   ☐ Transportation:

       ☐ Motorized boats   ☐ Non-motorized boats   ☐ Commercial

   ☐ Industrial water supply

   ☐ Waste disposal

   ☐ Other uses (explain) _____

3. Water use impairments:

   ☐ No   ☐ Yes. The impairment(s) is/are due to (check all that apply):

   ☐ Agricultural runoff   ☐ Livestock yards   ☐ Cropland/ Pasture

   ☐ Inadequate or overloaded wastewater treatment facilities:

       ☐ primary   ☐ secondary   ☐ tertiary

   ☐ Logging runoff   ☐ Mining runoff   ☐ Industrial Discharge

   ☐ Golf courses   ☐ Irrigation problems ☐ Housing

   ☐ Failing septic tanks ☐ Urban or other construction

   ☐ Other (explain) _____

DATA SHEET 6

WATER ODORS AND APPEARANCE

Odors

What smells does the watercourse have (check all that apply)?

		Water			*Soil*	
	Faint	Distinct	Strong	Faint	Distinct	Strong
Chemical:						
Chlorine	☐	☐	☐	☐	☐	☐
Sulfur (rotten eggs)	☐	☐	☐	☐	☐	☐
Musty:						
Decomposing straw	☐	☐	☐	☐	☐	☐
Moldy	☐	☐	☐	☐	☐	☐
Harsh:						
Fishy	☐	☐	☐	☐	☐	☐
Sewage	☐	☐	☐	☐	☐	☐
Earthy:						
Peaty	☐	☐	☐	☐	☐	☐
Grassy	☐	☐	☐	☐	☐	☐
Aromatic						
Spicy	☐	☐	☐	☐	☐	☐
Balsamic						
Flowery	☐	☐	☐	☐	☐	☐

☐ Other (explain) _____

☐ No unusual smells

Water Appearance

By visually observing the watercourse, what appearance does the water have (check all that apply)?

☐ Green   ☐ Orange-red   ☐ Foam   ☐ Reds

☐ Blues   ☐ Purples   ☐ Blacks   ☐ Milky/white

☐ Muddy/cloudy   ☐ Multi-colored (oily sheen)

☐ Other (explain) _____

☐ No unusual colors

INFORMATION SHEET

WATER ODORS AND APPEARANCE (2)

Water Appearance Color Standards

Verbal descriptions of apparent color can be unreliable and subjective. If possible, use a system of color comparison that is reproducible. By using established color standards, people in different areas can compare these results. Match your sample to a color standard. Record the reference number of the color standard yielding the best match. Be sure to report the system of color standards used along with your observations.

Two color standard systems:

➤ Forel Ule Color Scale—good for offshore and coastal bay waters.

➤ Borger Color System—good for natural waters, in addition, colors of insects, larvae, algae, and bacteria.

```
To order:  Ben Meadows Company
           3589 Broad Street
           Chamblee, GA 30341 USA
           USA and Canada:   1(800)241-6401
           Elsewhere:        1(404)455-0907
           Fax:              1(404)457-1841
```

Forel-Ule Color Scale Order #224220   Price US $51.00

Borger Color System Order #224218   Price US $5.95

DATA SHEET 7

HABITAT ASSESSMENT

1. Notice the places around you where plants and animals could live. Check all those items below that apply to the area of the catchment that you are observing.

   ☐ Pool   ☐ Riffle/rapids   ☐ Wetlands

   ☐ Rocks   ☐ Log piles   ☐ Weed beds

   ☐ Undercut banks

   ☐ Human-made objects (pilings, bridges) specify _____

   ☐ Other (please describe) _____

2. Animals are an important part of a catchment ecosystem. In the spaces below, list the name (if known) of all the fish, reptiles, and birds that were seen during your observations today. If the name is unknown, try to draw a picture that best depicts the animals that were seen.

   BIRDS

   FISH

   REPTILES

General Comments: _____

_____

_____

_____

3. In the box below, draw a cross-sectional area of the water under investigation. Include vegetation growth on banks, shape of channel, etc.

HABITAT ASSESSMENT   **SCORE** _____

   4 (excellent)   Bends present, 5–10 riffles in 10 meters, many snags
   3 (good)   Bends present, 1–2 riffles in 10 meters, some snags
   2 (fair)   Occasional bend, 1–2 riffles in 50 meters, few snags
   1 (poor)   Straight channel, riffles-pools absent, no snags

DATA SHEET 8

POLLUTION TOLERANCE INDEX (PTI)

Record the presence and estimate the number of each organism collected:

		Amount		
Group 1 (Index Value 4.0)	1–9	10–49	50–99	100 or more
Gill Snail	☐	☐	☐	☐
Stonefly	☐	☐	☐	☐
Mayfly	☐	☐	☐	☐
Riffle Beetle	☐	☐	☐	☐
Caddis Fly	☐	☐	☐	☐
Dobsonfly	☐	☐	☐	☐
Water Penny	☐	☐	☐	☐
Group 2 (Index Value 3.0)	1–9	10–49	50–99	100 or more
Sowbug	☐	☐	☐	☐
Scud	☐	☐	☐	☐
Dragonfly	☐	☐	☐	☐
Damselfly	☐	☐	☐	☐
Crane Fly	☐	☐	☐	☐
Clam	☐	☐	☐	☐
Group 3 (Index Value 2.0)	1–9	10–49	50–99	100 or more
Leech	☐	☐	☐	☐
Midge (excluding Blood Midges)	☐	☐	☐	☐
Flatworm	☐	☐	☐	☐
Black Fly	☐	☐	☐	☐
Water Mite	☐	☐	☐	☐
Group 4 (Index Value 1.0)	1–9	10–49	50–99	100 or more
Pouch Snail	☐	☐	☐	☐
Maggot	☐	☐	☐	☐
Tubifex	☐	☐	☐	☐
Blood Midge	☐	☐	☐	☐

CUMULATIVE INDEX VALUE   SCORE _____

4 (excellent)	23 or more
3 (good)	17–22
2 (fair)	11–16
1 (poor)	10 or less

DISTINGUISHING CHARACTERISTICS OF COMMON MACROINVERTEBRATE TAXA GROUPS

INDEX VALUE 4.0

These organisms are generally pollution-intolerant. Their dominance generally signifies good water quality.

Stonefly Nymph (Order Plecoptera):

- 5–35 mm long, not including tail; sometimes up to 60 mm.
- Eyes: Widely separated.
- Antennae: Long and slender.
- Body: 1st three segments are hard on top; dorsally-ventrally compressed (flat).
- Abdomen: Ends in 2 tail filaments (cerci).
- Gills: Sometimes lacking; if present, at the base of the legs.
- Legs: Well developed; each ends in 2 claws.

Mayfly Nymph (Order Ephemeroptera):

- 3–20 mm long (not including tail).
- Antennae: Slender and long.
- Legs: 3 pairs, well developed, segmented, single claw or no claw on end.
- Cylindrical to flattened shape; streamlined.
- Abdomen: Series (usually 7 pairs) of gills arising from side.
- Gills may have gill covers with feather-like appearance or may be flat and spade-shaped.
- Usually 3 tail filaments; occasionally 2.
- Two-developing fore-wing pads are evident.
- Movement: Side to side.

Caddisfly Larva (Order Trichoptera):

- 2–40 mm in length.
- Eyes: Small and simple.
- Head: Hard-shelled head capsule.
- Body: Segmented; 1st 3 segments behind head may have hard-shelled plates on top surface.
- Legs: 3 pairs of thoracic legs are well-developed.
- Abdomen: Soft and cylindrical; end of abdomen has a pair of hooks at end of abdomen usually hooked and sharp.
- Gills either underneath or at the side of the body.
- Larva may live within portable cases (made of stone, plant, material) or nets.
- Movement: Series of loops.

Dobsonfly Larva (Order Megaloptera):

- 25–90 mm.
- Head capsule is hardened.
- Abdomen: 8 pairs of lateral filaments extending from abdominal segments; each filament has 2 segments.

- No tail, one pair of anal prolegs, each with 2 terminal hooks.
- 1st three body segments are hardened.

Water Penney Larva (Order Coleoptera, Family Psephenidae):

- 1 cm long.
- Flattened and disc-like, almost as broad as long.
- Dorsal plate-like expansions conceal head and legs from above, almost as broad as they are long.
- Highly adapted for adhering to stones.

Gilled Snail (Family Lymnaeidae):

- Spiral-shaped cone.
- Right-handed shell spirals. (To determine spiral direction, hold the shell in the vertical orientation with the aperture facing you. If the aperture opens on the right and the shell spirals clockwise, the shell is dextral, or right-handed. A sinistral shell spirals counter-clockwise.)
- This snail has gills and obtains oxygen from the surrounding water.

INDEX VALUE 3.0

These organisms can exist in a wide range of water quality conditions.

Damselfly Nymph (Order Odonata, Suborder Zygoptera):

- 20–120 mm long.
- Slender.
- 2 pair of wing pads.
- Cylindrical abdomen.
- 3 well-developed long leaf-like appendages protruding from the back end, instead of tail filaments.

Dragonfly Nymph (Order Odonata, Suborder Anisoptera):

- 20–120 mm long.
- Slender.
- Abdomen: broadens from the base, becoming wider toward the back end.
- No tail; abdomen ends in 3 short, wedge-shaped structures.
- Mouth: Hinged, shovel-like lower jaw that can be extended remarkably.

Aquatic Sowbugs (Order Isopoda):

- 5–20 mm long.
- Shape: Flattened.
- Legs: 7 pairs; first two pairs are modified for grasping.
- Abdomen: Segments are fused into a relatively short region.

Scud (Order Amphipoda):

- 5–20 mm long.
- Shrimp-like; body flattened.
- Well developed eyes.

- Thorax: 7 segments.
- Legs: 7 pairs; first pair are modified for grasping.
- Abdomen: 6 segments.
- Movement: Swims backward, on its side or back.

Crane fly (Order Diptera, Family Tipulidae):

- 15–100 mm long.
- Head: Retractable, only partially hardened.
- Oblong, cylindrical, somewhat tapered toward head.
- Abdomen: Last segment has a six-lobed plate or 2–6 short finger-like lobes.

Freshwater Clam (Class Pelecypoda, Sub Family Pisisiidae):

- Bi-valve, two-piece shell.
- Shells usually oval, with concentric growth lines.

INDEX VALUE 2.0

These organisms are generally moderately tolerant of pollution. Their dominance usually signifies poor water quality.

Midge Larva (Order Diptera):

- 1 cm long.
- Legs: No jointed legs, as other true flies have.
- 2 pairs of pro-legs (fleshy and not jointed, short and stumpy), 1 pair just below head, 1 pair at back end.
- Body: Soft, slender, cylindrical; almost always curved in "C" or "S" shape.
- Color: Some common species are blood red (have hemoglobin which enables them to survive in water with little dissolved oxygen), others are not red.

Black fly Larva (Order Diptera, Family Simuliidae):

- Small size.
- Normally attached by its rear end to a substrate.
- Abdomen: Posterior part of the abdomen noticeably swollen.

Flatworm—Planaria (Order Tricladida, Family Planariidae):

- 1–30 mm long.
- Flattened bodies, elongate bodies
- Generally white, gray brown, or black
- Up to 30 mm
- They glide over submerged plants or stones
- Acutely triangular head

Leeches (Class Hirudinea):

- 5–400 mm long.
- Body: Many segmented; appears flattened dorsally-ventrally (top-to-bottom).
- Sucker on both ends of the body.
- May be patterned or brightly colored.
- Movement: By loops.

Water Mite (Class Hydracarina)

- Small spider-like animals
- Eight legs
- Some have a red pigment.
- Swim freely in water

INDEX VALUE 1.0

These organisms are very tolerant of pollution. Their dominance usually signifies very poor water quality

Pouch or Lung Snails (Order Prosobranchia, Family Physidae):

- Tolerant of poor water conditions; they have lungs and come to the surface to breathe.
- Taller than they are wide.
- Pouch snails are sinistral or left-handed (see Physidae in Group 1 for explanation).
- Pouch snails do not have hemoglobin.

Sewage Worms (Family Tubificidae):

- 5 cm to 8 cm long
- Worm like and tapered, with sheath around base of body
- Body: soft, slender, cylindrical.

Blood Midge (Order Diptera):

- 1 cm. long.
- Legs: No jointed legs, as other true flies have.
- 2 pairs of pro-legs (fleshy and not jointed, short and stumpy), 1 pair just below head, 1 pair at back end.
- Body: Soft, slender, cylindrical; almost always curved in "C" or "S" shape.
- Color: Blood red (have hemoglobin which enables them to survive in water
- with little dissolved oxygen.

Rat-tailed Maggot (Family Syrphidae)

- Long posterior respiratory tube from once to several times the length of the body.
- Can live in polluted water, drawing air from the water surface through their respiratory tube.

DATA SHEET 9

DISSOLVED OXYGEN

Date: _____

Testing Site Location: _____

Time: _____

Weather Conditions: _____

Names of team members:

_____   _____

_____   _____

_____   _____

_____   _____

Temperature Reading = _____ (in °C)

_____ mg/liter → _____ % Saturation (from chart).

_____ mg/liter → _____ % Saturation (from chart).

_____ mg/liter → _____ % Saturation (from chart).

_____ mg/liter → _____ % Saturation (from chart).

ADD UP ALL OF THE REASONABLE VALUES AND DIVIDE BY THE
NUMBER OF SAMPLES, (I.E. TAKE THE AVERAGE), TO GET THE
OFFICIAL MG/LITER (PPM) AND SATURATION LEVEL FOR YOUR SITE.

Dissolved Oxygen= _____mg/liter

% Saturation= _____

DATA SHEET 10

FECAL COLIFORM

Date: _____

Testing Site Location: _____

Time: _____

Weather Conditions: _____

Name of team members:

_____   _____

_____   _____

_____   _____

_____   _____

REMEMBER THAT ALL SAMPLE BOTTLES, PIPETTES, AND FILTRATION
SYSTEMS MUST BE STERILIZED BEFORE SAMPLING

1) Volume of sample _____(mL)

2)   → # _____ colonies after incubation

3)   conversion necessary → _____ Colonies/100 mL

1) Volume of sample _____(mL)

2)   → # _____ colonies after incubation

3)   conversion necessary → _____ Colonies/100 mL

1) Volume of sample _____(mL)

2)   → # _____ colonies after incubation

3)   conversion necessary → _____ Colonies/100 mL

IT IS IMPORTANT THAT YOU REPORT THE HIGHEST FECAL
COLIFORM VALUE RATHER THAN THE AVERAGE.

Official Reading =_____Colonies/100 mL

DATA SHEET 11

pH

Date: _____

Testing Site Location: _____

Time: _____

Weather Conditions: _____

Names of team members:

_____   _____

_____   _____

_____   _____

_____   _____

_____   _____

YOU MUST RUN THE TEST FOR pH IMMEDIATELY AFTER
SAMPLING BECAUSE CHANGES IN TEMPERATURE OF THE
SAMPLE CAN CHANGE THE MEASURED pH.

Values for test repetitions, (Try to take at least three measurements):

_____   _____   _____   _____

_____   _____   _____   _____

Take the most common value (mode) to report.

Official pH Reading = _____

DATA SHEET 12
BOD

Date: _____

Testing Site Location: _____

Time: _____

Weather Conditions: _____

Names of team members:

_____   _____

_____   _____

_____   _____

_____   _____

Temperature Reading = _____ (in °C)

First DO Test Results:

_____ mg/liter → _____ % Saturation (from chart).

_____ mg/liter → _____ % Saturation (from chart).

TAKE THE AVERAGE TO GET THE MG/LITER (PPM)
AND SATURATION LEVEL.

Dissolved Oxygen = _____ mg/liter   % Saturation = _____

DO Test Results AFTER FIVE DAYS

_____ mg/liter → _____ % Saturation (from chart).

_____ mg/liter → _____ % Saturation (from chart).

TAKE THE AVERAGE TO GET THE MG/LITER (PPM)
AND SATURATION LEVEL.

Dissolved Oxygen = _____ mg/liter   % Saturation = _____

BOD= _____ mg/liter (first reading)   *minus* _____ mg/liter (after 5 days) =

_____ mg/liter

DATA SHEET 13

TEMPERATURE

Date: _____

Testing Site Location: _____

Time: _____

Weather Conditions: _____

Names of team members:

_____   _____

_____   _____

_____   _____

_____   _____

REMEMBER: TEMPERATURE READINGS SHOULD BE IN °C.
TRY TO TAKE AT LEAST THREE SETS OF TEMPERATURE READINGS.

_____ Temp. downstream – _____ Temp. upstream = _____

_____ Temp. downstream – _____ Temp. upstream = _____

_____ Temp. downstream – _____ Temp. upstream = _____

_____ Temp. downstream – _____ Temp. upstream = _____

_____ Temp. downstream – _____ Temp. upstream = _____

_____ Temp. downstream – _____ Temp. upstream = _____

_____ Temp. downstream – _____ Temp. upstream = _____

TAKE THE MOST COMMON VALUE (MODE) TO REPORT.

Δ Temperature = _____

DATA SHEET 14

TOTAL PHOSPHATE

Date: _____

Testing Site Location: _____

Time: _____

Weather Conditions: _____

Names of team members:

_____   _____

_____   _____

_____   _____

_____   _____

_____ mg/liter Phosphate

_____ mg/liter Phosphate

_____ mg/liter Phosphate

TAKE AVERAGE IF YOU TAKE MORE THAN ONE READING.

Official Total Phosphate Reading = _____mg/liter

DATA SHEET 15

NITRATES

Date: _____

Testing Site Location: _____

Time: _____

Weather Conditions: _____

Names of team members:

_____   _____

_____   _____

_____   _____

_____   _____

REMEMBER TO DISPOSE OF THE TOXIC WASTE PRODUCTS
OF THIS TEST APPROPRIATELY.

_____ Reading x 4.4 = mg/liter Nitrate

_____ Reading x 4.4 = mg/liter Nitrate

_____ Reading x 4.4 = mg/liter Nitrate

TAKE THE AVERAGE IF YOU TAKE MORE THAN ONE READING.

Official Nitrate reading = _____mg/liter

DATA SHEET 16

TURBIDITY

Date: _____

Testing Site Location: _____

Time: _____

Weather Conditions: _____

Names of team members:

_____   _____

_____   _____

_____   _____

_____   _____

IF USING A SECCHI DISK (MEASUREMENTS IN FEET):

_____ (depth the Secchi disk disappears + depth disk reappears) ÷ 2 = _____feet

_____ (depth the Secchi disk disappears + depth disk reappears) ÷ 2 = _____feet

_____ (depth the Secchi disk disappears + depth disk reappears) ÷ 2 = _____feet

TAKE THE MOST COMMON VALUE (MODE) TO REPORT.

Official Reading = _____ Feet Turbidity

DATA SHEET 17

TOTAL SOLIDS

Date: _____

Testing Site Location: _____

Time: _____

Weather Conditions: _____

Names of team members:

_____   _____

_____   _____

_____   _____

_____   _____

Weight of 300mL beaker WITH RESIDUE = _____ grams

– Weight of empty 300mL beaker BEFORE drying = _____ grams

Weight of residue =_____grams

Formula for determining total solids is:

$$\frac{\text{Weight of residue (in grams)}}{\text{Volume of sample (in mL)}} \times \frac{1000 \text{ mg}}{1 \text{ gram}} \times \frac{1000 \text{ mL}}{1 \text{ liter}} = \underline{\qquad} \text{ mg/liter}$$

Official Reading = _____ mg/liter

CHEMICAL REACTION EQUATIONS FOR THE CHEMICAL TESTS

Dissolved Oxygen Test

Dissolved oxygen is often tested using the Azide modification of the Winkler method. When testing dissolved oxygen it is critical not to introduce additional oxygen into the sample. Many people avoid this problem by filling the sample bottle all the way and allowing the water to overflow for one minute before capping.

The first step in a DO titration is the addition of Manganous Sulfate Solution (4167) and Alkaline Potassium Iodide Azide Solution (7166). These reagents react to form a white precipitate, or floc, of manganous hydroxide, $Mn(OH)_2$. Chemically, this reaction can be written as:

$$MnSO_4 + 2KOH \rightarrow Mn(OH)_2 + K_2SO_4$$

Manganous Sulfate + Potassium Hydroxide → Manganous Hydroxide + Potassium Sulfate

Immediately upon formation of the precipitate, the oxygen in the water oxidizes an equivalent amount of the manganous hydroxide to brown-colored manganic hydroxide. For every molecule of oxygen in the water, four molecules of manganous hydroxide is converted to manganic hydroxide. Chemically, this reaction can be written as:

$$4Mn(OH)_2 + O_2 + 2H_2O \rightarrow 4Mn(OH)_3$$

Manganous Hydroxide + Oxygen + Water → Manganic Hydroxide

After the brown precipitate is formed, a strong acid, such as Sulfamic Acid Powder (6286) or Sulfuric Acid, 1:1 (6141) is added to the sample. The acid converts the manganic hydroxide to manganic sulfate. At this point the sample is considered "fixed" and concern for additional oxygen being introduced into the sample is reduced. Chemically, this reaction can be written as:

$$2Mn(OH)_3 + 3H_2SO_4 \rightarrow Mn_2(SO_4)_3 + 6H_2O$$

Manganic Hydroxide + Sulfuric Acid → Manganic Sulfate + Water

Simultaneously, iodine from the potassium iodide in the Alkaline Potassium Iodide Azide Solution is oxidized by manganic sulfate, releasing free iodine into the water. Since the manganic sulfate for this reaction comes from the reaction between the manganous hydroxide and oxygen, the amount of iodine released is directly proportional to the amount of oxygen present in the original sample. The release of free iodine is indicated by the sample turning a yellow-brown color. Chemically, this reaction can be written as:

$$Mn_2(SO_4)_3 + 2KI \rightarrow 2MnSO_4 + K_2SO_4 + I_2$$

Manganic Sulfate + Potassium Iodide → Manganous Sulfate + Potassium Sulfate + Iodine

The final stage in the Winkler titration is the addition of sodium thiosulfate. The sodium thiosulfate reacts with the free iodine to produce sodium iodide. When all the iodine has been converted the sample changes from yellow-brown to colorless. Often a starch indicator is added to enhance the final endpoint. Chemically, this reaction can be written as:

$$2Na_2S_2O_3 + I_2 \rightarrow Na_2S_4O_6 + 2NaI$$

Sodium Thiosulfate + Iodine → Sodium Tetrathionate + Sodium Iodide

CHEMICAL REACTION EQUATIONS FOR THE CHEMICAL TESTS (2)

pH test

$$HAc + H_2O \rightarrow Ac^- + H_3O^+$$
$$B + H_2O \rightarrow BH + OH^-$$

Acids react with water to form hydronium ions, H_3O^+. A reaction with water is called hydrolysis. Bases, B, hydrolyze water to form hydroxide ions, OH^-. When there are more hydronium ions than hydroxide ions, the water has a pH less than 7 and is therefore acidic. When there are more hydroxide ions than hydronium ions, the pH is greater than 7 and is therefore basic.

$$H_3O^+ + +In^- \rightarrow HIn + H_2O$$
$$OH^- + HIn \rightarrow In^- + H_2O$$

pH indicators, HIn, are weak acids or bases which change colors depending on the pH. HIn, the acid form of the indicator, has one color, while In^-, the basic form, has another.

Nitrate test

The test for nitrate involves an oxidation-reduction reaction in one molecule loses electrons, gets oxidized, and another molecule gains electrons, gets reduced. Cadmium, Cd, is oxidized and reduces nitrate, NO_3^-, to nitrate, NO_2^-. Nitrate then reacts with a nitrite indicator, In, to form a pink dye. The intensity of the color is proportional to the amount of nitrate originally present.

$$NO_3^- + Cd + 2H^+ \rightarrow NO_2^- + H_2O + Cd^2_+$$
$$NO_2^- + In \rightarrow Dye$$

Phosphate

Phosphate exists in water as part of organic molecules, inorganic polyphosphates and as ortho phosphates, PO_4^{3-}. To test for all forms of phosphate they must be treated to a digestion which converts all forms to orthophosphate. Orthophosphate reacts with molybdate to form a molybdophosphate. The molybdophosphate is then reduced by ascorbic acid to form a molybdophosphate blue complex. The intensity of the blue color, which results, is proportional to the total phosphate that was present in a sample.

$$\text{organic, inorganic, and orthophosphates} \xrightarrow{acid} PO_4^{3-}$$
$$PO_4^{3-} + (NH4)MoO4 \rightarrow (NH_4)_3[P(MO_3O_{10})_4]$$
$$(NH_4)_3[P(MO_3O_{10})_4] + \text{ascorbic acid} \rightarrow (MoO_2 \bullet 4MoO_3)_2 \bullet H_3PO_4$$
$$\textit{molybdophosphate blue}$$

WATER QUALITY MONITORING QUANTITATIVE ANALYSIS

VERGE VEGETATION   SCORE _____

4 (excellent)	Vegetation present and canopy intact
3 (good)	Vegetation and canopy virtually intact
2 (fair)	Vegetation clearly disturbed
1 (poor)	Cleared land or urban development

BANK VEGETATION   SCORE _____

4 (excellent)	Vegetation in undisturbed state
3 (good)	Vegetation mildly disturbed
2 (fair)	Vegetation moderately disturbed
1 (poor)	Vegetation severely disturbed

BANK BARE SOIL (PERCENT)   SCORE _____

4 (excellent)	0–10
3 (good)	11–40
2 (fair)	41–80
1 (poor)	81–100

BANK EROSION   SCORE _____

4 (excellent)	Stable, no sign of any bank erosion
3 (good)	Very occasional and local erosion
2 (fair)	Some erosion evident
1 (poor)	Severe bank failure with extensive cracking and fall-ins

BANK SLUMPING AND MOVEMENT   SCORE _____

4 (excellent)	No movement
3 (good)	Slight movement on the bank
2 (fair)	Moderate bank collapses
1 (poor)	Severe bank failure with extensive cracking and fall-ins

BENDS AND RIFFLES   SCORE _____

4 (excellent)	Bends present, 5–10 riffles in 10 meters, many snags
3 (good)	Bends present, 1–2 riffles in 10 meters, some snags
2 (fair)	Occasional bend, 1–2 riffles in 50 meters, few snags
1 (poor)	Straight channel, riffles/pools absent, no snags

PHYTOPLANKTON   SCORE _____

4 (excellent)	High diversity of phytoplankton (blue-green, green diatoms, flagellates)
3 (good)	Alert level I (500–2000 potentially toxic blue-green algal cells/mL)
2 (fair)	Alert level II (2000–15,000 potentially toxic blue-green algal cells/mL) concern for drinking water supplies
1 (poor)	Alert level III (greater than 15,000 potentially toxic blue-green algal cells/mL) can cause death and illness in cattle and humans

MACROPHYTE RIVER COVER   SCORE _____

4 (excellent)   Patches of surface and underwater plant cover (<10%), abundant overhanging vegetation

3 (good)   Some surface and underwater plant cover (10–30%), some overhanging vegetation

2 (fair)   Abundant surface and underwater plant cover (10–50%), little overhanging vegetation

1 (poor)   Choked surface and underwater plant cover (50–100%), no overhanging vegetation

SEQUENTIAL COMPARISON INDEX   SCORE _____
(Benthic Macroinvertebrates)

4 (excellent)   0.9–1.0
3 (good)   0.6–0.89
2 (fair)   0.3–0.59
1 (poor)   0.0–0.29

POLLUTION TOLERANCE INDEX   SCORE _____
(Benthic Macroinvertebrates)

4 (excellent)   23 and above
3 (good)   17–22
2 (fair)   11–16
1 (poor)   10 or less

EPHEMEROPTERA, PLECOPTERA, TRICOPTERA   SCORE _____
(Benthic Macroinvertebrates)

4 (excellent)   more than 15 families
3 (good)   12–15 families
2 (fair)   8–12 families
1 (poor)   less than 8 families

DISSOLVED OXYGEN LEVELS (% SATURATION)   SCORE _____

4 (excellent)   91–110
3 (good)   71–90
2 (fair)   51–70
1 (poor)   <50

FECAL COLIFORM (PER 100 ML)   SCORE _____

4 (excellent)   <50 colonies
3 (good)   51–200 colonies
2 (fair)   100–1,000 colonies
1 (poor)   >1,000 colonies

pH (UNITS) SCORE _____
    4 (excellent)   6.5–7.5
    3 (good)   6.0–6.4, 7.6–8.0
    2 (fair)   5.5–5.9, 8.1–8.5
    1 (poor)   <5.5, >8.6

BIOCHEMICAL OXYGEN DEMAND (PPM) SCORE _____
    4 (excellent)   <2
    3 (good)   2–4
    2 (fair)   4.1–10
    1 (poor)   >10

TEMPERATURE CHANGE (DEGREES CELSIUS) SCORE _____
    4 (excellent)   0–2
    3 (good)   2.2–5
    2 (fair)   5.1–9.9
    1 (poor)   10>

TOTAL PHOSPHATE (MG/L) SCORE _____
    4 (excellent)   0–1
    3 (good)   1.1–4
    2 (fair)   4.1–9.9
    1 (poor)   >10

NITRATES (MG/L) SCORE _____
    4 (excellent)   0–1
    3 (good)   1.1–3 (good)
    2 (fair)   3.1–5
    1 (poor)   >5

TURBIDITY (NTUS/FEET OR CM) SCORE _____

	NTUs	Secchi Disk
4 (excellent)	0–10	>3 feet
		>91.5 cm
3 (good)	10.1–40	1 foot to 3 feet
		30.5 cm to 91.5 cm
2 (fair)	40.1–150	2 inches to 1 foot
		5 cm to 30.5 cm
1 (poor)	>150	<2 inches
		<5 cm

TOTAL SOLIDS (MG/L) SCORE _____
    4 (excellent)   <100
    3 (good)   100–250
    2 (fair)   250–400
    1 (poor)   >400

Appendix C
References

American Public Health Association. 1990. *Standard Methods for the Examination of Water and Wastewater.* 18th ed. American Public Health Association, Inc., New York.

Aspen Global Change Institute. 1992. *Ground Truth Studies Teacher Handbook.* 100 East Francis Street, Aspen, Colorado, 81611.

Caduto, M.J. 1990. *Pond and brook: a guide to nature study in freshwater environments.* 2nd ed. Prentice-Hall, N.J.

Cairns, and K. L. Dickson. 1971. "A simple method for the biological assessment of the effects of waste discharges on aquatic bottom dwelling organisms," *Journal of Water Pollution Control Fed.* 43: 755-771.

Cromwell, Mare et al. 1992. *Investigating Streams and Rivers.* GREEN, *University of Michigan, Ann Arbor.*

Cummins, K.W. 1979. "From headwater streams to rivers." *American Biology Teacher.* 34: 305–312.

Fergusson, Jack E. l990. *The heavy elements: chemistry, environmental impact, and health effects.* Pergamon Press.

Forstner, U. 1979. *Metal Pollution in the Aquatic Ecosystem.* SpringerVerlag, Berlin: New York, 1979.

Friberg, Lars; Nordberg, Gunnar, F.; & Velimir B. Vouk. 1986. *Handbook on the Toxicology of Metals.* Elsevier Pub.

Great Lakes Water Quality Board, International Joint Commission. 1987. "Restoration of Areas of Concern" in *Report on Great Lakes Water Quality—A Report to the International Joint Commission from the Great Lakes Water Quality Board.*

Hynes, H.B.N. 1974. *The Biology of Polluted Waters.* University of Toronto Press, Toronto. (202 pp.)

—. 1970. *The Ecology of Running Waters.* University of Toronto Press, Toronto. (555 pp.)

Klots, Elsie B. 1966. *The New Field Book of Freshwater Life.* G.P. Putnam's Sons, New York.

Kopec, J. and S. Lewis. 1989. *Stream Monitoring: A Citizen Action Program.* Ohio Department of Natural Resources, Division of Natural Areas and Preserves; Scenic Rivers Program.

Lee, V. and Eleanor Ely. 1990. *National Directory of Citizen Volunteer Environmental Monitoring Programs.* U.S. Environmental Protection Agency and the University of Rhode Island.

McCafferty, P.W. 1981. *Aquaticentomology: the fishermen's and ecologist's guide to insects and their relatives.* Jones and Bartlett Publishers, Inc. California.

Merritt, R.E. and K.W. Cummins. 1988. *An Introduction to the Aquatic Insects of North America.*

Needham, James G. and Paul Needham. 1962. *A Guide to the Study of Freshwater Biology.* Holden-Day, Inc., San Francisco.

Pennaok, Robert. 1973. *Freshwater Invertebrates of the United States.* Ronald Press, NY.

Reid, G.K. 1987. *Pond Life* (A Golden Guide), Golden Press, N.Y.

River Watch Network. 1990. *Guide to Macroinvertebrate Sampling for White River Headwaters Citizens Monitoring Group.* (Adapted from: *Laypersons Guide to Stream or River Water Quality—Biological Monitoring* by Steve Fiske, Vermont Department of Water Resources, 1978.) Montpelier, Vermont.

Terrill, C.R. and P. B. Perfetti. 1988. *Water Quality Indicators Guide: Surface Waters.* USOA, SCS-TP-161.

U.S. Environmental Protection Agency: Office of River Resources. 1991. *Manual for Citizen Volunteers—River Monitoring.* Washington, D.C.

Ward, Robert and W. Whipple. 1959. *Freshwater Biology.* John Wiley & Sons, NY.

Whipple, W., Jr. And J.V. Hunter. 1977. "Nonpoint sources and planning for water pollution control," *Journal of Water Pollution Control Fed.* 49: 15–23.

Yates, S. 1991. *Adopting a stream: A northwest handbook.* Adopt-a-Stream Foundation. Published and distributed by University of Washington Press.

River Watch Network
153 State St.
Montpelier, Vermont 05602

Save Our Streams
The Izaak Walton League of America
1800 North Kent Street, Suite 806
Arlington, Virginia 22209

Index

GREEN
Global Rivers Environmental Education Network
Global citizens sharing their concerns for water quality

GREEN is an innovative, action-oriented approach to education, based on an interdisciplinary watershed education model. GREEN's mission is to improve education through a global network that promotes watershed sustainability. It is a resource to schools and communities that wish to study their watershed and work to improve their quality of life.

GREEN works closely with educational and environmental organizations, community groups, businesses, and government across the United States and Canada, and in over 130 countries around the globe to support local efforts in watershed education and sustainability.

> Students at North Farmington High School near Detroit detected elevated levels of bacterial contamination down river from a pipe exiting a City sewage pumping station. They presented their findings to the City Engineer, who acted quickly to repair the malfunctioning pump.

GREEN Watershed Education Model

The model involves the synthesis of content and process. Activities revolve around two key areas: watershed and water quality monitoring, and understanding changes and trends within the whole watershed.

GREEN participants collect and analyze real-life environmental data; study current and historical patterns of land and water usage within their watershed; share their data, concerns and strategies for action with others in the watershed and beyond; and develop concrete action plans to improve local water quality.

Key to the GREEN process is an emphasis on creating a learning community of teachers, students, parents, community groups, government, nongovernmental organizations, and businesses—whose members share a vision for watershed sustainability and possess the skills, knowledge, and motivation necessary to create change.

GREEN Workshops and Institutes

GREEN can help you start or enhance your watershed education program. We'll create and deliver a **Custom Workshop** with support materials to fit the goals of your program and participants. Topics include:

➤ Starting a Watershed Education Program

➤ Interdisciplinary Approaches to Watershed Education

➤ Water Quality Monitoring & Data Interpretation

➤ Land Use Analysis With Maps & Satellite Imagery

➤ Telecommunications in Watershed Education

➤ Student Problem Solving & Action Taking

All workshops provide hands-on experience in the GREEN approach for watershed education; a variety of support materials; a basic understanding of the complex relationships between land, water, and people; and activities that demonstrate the importance of protecting the environment and how individuals and communities can effect environmental change.

GREEN also hosts a series of regularly scheduled, introductory **Watershed Education Institutes** at locations across the U.S. and Canada. Please contact GREEN for details.

A local business man working with Project del Rio, an international GREEN program that links schools along the U.S.-Mexico border, is convinced that the GREEN model is "exactly the kind of program that enables a community to participate in the preservation and restoration of its environment."

On the Cutting Edge

GREEN is engaged in numerous initiatives to develop and refine curricular materials and to pilot **innovative approaches to education** such as school-community linkages, inquiry-based learning, action research, use of telecommunications and technology, and cross cultural learning. With our partners we are developing software for modeling environmental data, digital technology for field investigations, Internet-accessible environmental database technology, and low-cost methods for environmental monitoring.

A fundamental part of our research and development strategy is the suite of GREEN publications. GREEN publishes manuals, curriculum guides, and videos to support global watershed education. Our titles cover topics such as: water quality monitoring, action taking, and cross-cultural partnerships. Contact Earth Force if you would like to receive our catalog of educational materials and water monitoring equipment.

GREEN Publications

- **Field Manual for Water Quality Monitoring: An Environmental Education Program for Schools, Twelfth Edition,** details nine chemical/physical water quality tests and methods for biological monitoring. It also includes chapters on heavy metals testing, land use practices, action taking, and computer networking. The Field Manual is the foundation for GREEN's Educational Model.

  The following books provide more in-depth information on components of their model.

- *Field Manual for Global Low-Cost Water Quality Monitoring, Second Edition,* provides a global perspective for watershed education. It includes activities to help readers understand key concepts and build skills. It provides handouts and instructions for making inexpensive equipment.

- *Investigating Streams and Rivers* is an activity guide that promotes an interdisciplinary approach to watershed education. It focuses on action taking and enhancing student involvement through computer networks.

- *Sourcebook for Watershed Education* provides detailed guidelines for the development of watershed-wide education programs, focusing on program goals, funding, and school-community partnerships. It contains a rich set of interdisciplinary classroom activities and outlines GREEN's educational philosophy.

- *Cross Cultural Watershed Partners: Activities Manual* contains activities for use in an intercultural watershed education program, and suggestions on how to structure a cross cultural exchange around watershed themes.
- *Air Pollution: Ozone Study and Action* moves students from awareness of air pollution and ozone to a point where they will be knowledgeable and empowered to make action to address problems in their own lives and communities.
- *Environmental Education for Empowerment* enables students, teachers, administrators, and others to effectively participate in the planning, implementation and evaluation of educational activities aimed at resolving an environmental issue that they themselves have identified.
- *International Case Studies on Watershed Education* provides a rich picture of the kaleidoscope of programs world-wide. These case studies sensitize the reader to potential implementation barriers and offer a vast number of new ideas and resources for school and community based programs.

About GREEN

The impetus for GREEN began the spring of 1984 with a group of concerned students at a high school located along the polluted Huron River in Ann Arbor, Michigan. Their teacher contacted Dr. William Stapp and other educators at the University of Michigan, and together they developed a comprehensive educational program called GREEN.

The idea quickly caught on; experiences gained in three years of work with schools along the Huron set the stage for an expanded program on the Rouge River in 1987—part of an effort to improve education and the environment in the broader Detroit metropolitan area.

The educational model moved to other watersheds around the Great Lakes in the U.S. and Canada. As the program expanded nationally and then internationally, other components were added: community partnerships, computer telecommunications, cross cultural opportunities, and integration of GREEN's initiatives across the curriculum to form a comprehensive program for watershed sustainability.

GREEN is a program under the umbrella of Earth Force. Earth Force is a national, non profit organization committed to young people changing their communities while developing life-long habits of active citizenship and environmental stewardship.

You can contact GREEN at:

Earth Force
1908 Mt. Vernon Ave., 2nd Flr
Alexandria, VA 22301

Tel: (703) 299-9400
Fax: (703) 299-9485

Internet: www.earthforce.org
Email: green@earthforce.org

YES!!

send me

Environmental Education Books
by Earth Force's Global Rivers
Environmental Education Network

Qty	Book Author/Title	Price	Total
	Cole-Misch, et al (GREEN) 1-2372 Sourcebook for Watershed Education	$29.95*	
	Frank et al (GREEN) 1-2380 Air Pollution: Ozone Study and Action	$19.95*	
	Mitchell-Stapp (GREEN) 1-2345 Field Manual for Water Quality Monitoring: An Environmental Education Program for Schools, Eleventh Edition	$19.95*	
	Stapp, et al (GREEN) 1-2390 Cross Cultural Watershed Partners	$14.95*	
	Stapp, et al (GREEN) 1-2341 Environmental Education for Empowerment	$19.95*	
	Stapp, et al (GREEN) 1-2385 Investigating Streams and Rivers	$14.95*	
	Stapp-Mitchell (GREEN) 1-2375 Field Manual for Global Low-Cost Water Quality Monitoring, Second Edition	$19.95*	
	Stapp, et al (GREEN) 1-3027 International Case Studies on Watershed Education	$19.95*	

Method of payment:

☐ check enclosed   ☐ charge my account

☐ Mastercard   ☐ VISA   ☐ American Express

Card Number _____

Expiration Date _____

Signature (required) _____

Name _____

Affiliation _____

Address _____

City _____

State/Zip _____

Phone _____

*Price subject to change without notice.

Call 1-800-228-0810 or fax 1-800-772-9165

KENDALL/HUNT PUBLISHING COMPANY
4050 Westmark Drive • P.O. Box 1840 • Dubuque, IA 52004-1840